Unless Recalled Earlier
DATE DUE

DEMCO, INC. 38-2931

The Science Frontier Express Series

THE ROAD TO CHAOS

Yoshisuke Ueda
Department of Electrical Engineering
Kyoto University, Japan

with the assistance of

Ralph H. Abraham
University of California at Santa Cruz

and

H. Bruce Stewart
Brookhaven National Laboratory

Aerial Press, Inc.
P. O. Box 1360, Santa Cruz, CA 95061

Library of Congress Catalog Card Number: 92-73630
ISBN 0-942344-14-6

PREFACE

Aerial Press is proud to present this volume of selected papers by Professor Yoshi Ueda. We actively promoted this project, and persuaded Professor Ueda to help. Bruce Stewart and I named the book, expressing our conviction of the important role of Ueda's work in the creation of chaos theory. So, perhaps a personal word is in order here on the tradition of experimental dynamics in Japan, and its role in my own development.

Let me begin with Ueda's teacher, Chihiro Hayashi (1911-1987). He had studied the works of Lord Rayleigh, Poincaré, Liapunov, Duffing, Van der Pol, and the other classics of dynamics. In an environment of electronic engineering, he evolved his own style in the period 1942-1948, a unique combination of electronic experiments in the style of Van der Pol, classical analysis following Poincaré, and a new level of excellence in graphical presentation. This work appeared in his book in English in 1953 (republished in a much expanded version in 1964 and in 1985), and began to have an influence in Europe and the USA after his visit to MIT in 1955-56. My own discovery of Hayashi's book came relatively late, in the mid-1970s, just as I was beginning the development of the Visual Math Project at the University of California at Santa Cruz, and I must acknowledge the powerful influence of Hayashi's graphics in the visual element of my own work. Many results and illustrations added to Hayashi's second book of 1964 were made by his students: H. Shibayama, M. Kuwahara, Y. Nishikawa, M. Abe and Y. Ueda. In 1984 I went on pilgrimage to Kyoto to meet Hayashi, who treated me very well. At this time Hayashi introduced me to several of his former students, and thus, I met Ueda, whose work was already well-known to our group in Santa Cruz.

At that time I did not know the story, told at last in this book, of the difficult birth of the chaos revolution in Ueda's work in 1961 in Kyoto. But by now it is clear that crucial steps on the road to chaos were taken first in Kyoto, and are still diffusing outward. The exact moment of this bifurcation is now known, as you will see in the Comments by Bruce Stewart at the start of the book, and by Ueda himself at the end. Hayashi was the last master of the periodic paradigm, and Ueda the first master of the chaotic. The following chronology will help in following the steps to chaos described in this book.

1942	Hayashi begins his original research program on forced oscillators
1953	First edition of Hayashi's book
1959	Ueda begins doctoral research with Hayashi on forced oscillations
1961	Ueda observes chaos in analog simulation of forced oscillations of the mixed Van der Pol/Duffing system (Selection 1, Chapter 4, Fig. 4.7; Selection 7, Fig. 1)
1963	Ueda observes chaos in analog simulation of the Duffing system (Selection 7, Chapter 3)
1964	Second edition of Hayashi's book with additional data and illustrations by his students
1965	Ueda finishes thesis (Selection 1)
1969	Ueda submits his work on chaotic attractors for publication (Selection 2)
1973	Ueda's second paper on chaos published (Selection 3)
1981	Hayashi presents his first paper on chaos in Kiev

It is a great pleasure to introduce this important book, which reports the bifurcation of the graphic experimental tradition of Japan from the periodic epoch of Hayashi into the chaotic era of Ueda, in the first three Selections, and the other outstanding accomplishments of Professor Ueda since those early days. The international community of dynamical systems specialists still have much to learn from this body of work, spanning the past thirty years, and presented here in an accessible single volume for the first time.

Ralph Abraham
Santa Cruz

TECHNICAL COMMENTS

Since the first appearance of the "Picture book" [16]* in 1980, a wide audience has learned about chaos and nonlinear behavior in Duffing's equation and other forced oscillators through Yoshisuke Ueda's research papers. His earlier research is not so widely known, but deserves consideration. Like his contemporary Lorenz, he charted new territory, and those who follow can benefit from another look at the origins of the chaos paradigm.

Chaos was first noticed and recorded by Ueda at the end of 1961 as part of his doctoral thesis research on forced oscillators under the guidance of Chihiro Hayashi. Ueda had determined the approximate regions of entrainment of a forced self-oscillator by harmonic balance methods. As Kyoto University did not yet have a general use digital computer, his predictions could at first only be tested by analog simulation. What Ueda found outside the regions of entrainment did not fit into the then-known universe of nonlinear oscillation theory. By the time the thesis was completed in 1965, chaos had appeared in the Duffing equation as well, but Hayashi saw no reason to look beyond the accepted concept of almost periodic oscillation, so the issue was not joined in the thesis [3], nor in the 1968 monograph condensed from it [5].

By 1969 Ueda was persuaded that a previously unheralded type of long-term or steady oscillation existed, which needed further attention. He submitted the manuscript, published in 1970 [7], giving examples of what are now called a chaotic attractor and a fractal basin boundary.

In 1973, Ueda published a paper [10] setting forth his understanding of chaos, following a close reading of G. D. Birkhoff's theory of recurrence and central motions.

These first three selections have been revised for this collection, with the purpose of improving the English translation. Care has been taken to avoid any changes which would be out of keeping with Ueda's understanding at the time the papers were first written. A few phrases and sentences have been added for clarity or emphasis; these additions are often repetitions or minor variations of points made elsewhere in the same work.

Doctoral Thesis   This text is condensed, and includes revisions made to Chapter 3 for the 1968 monograph.

---

*Numbers in brackets refer to the List of Publications in English at the end of this volume.

It should be noted that the term beat oscillation has been substituted in Chapters 4 and 5 wherever the thesis misleadingly used the term almost periodic oscillation to describe chaotic behavior.

Figure 1.2 has a particularly interesting history. A similar figure with $k = 0.2$, computed and drawn by Ueda, will be found on page 129 of Hayashi's 1964 book "Nonlinear Oscillations in Physical Systems." The second unstable region in both figures was a source of misunderstanding between Hayashi and Ueda. For $k = 0.2$, this region is predominated by chaotic attractors, while for $k = 0.4$ chaotic transients are typical; see also Fig. 7 of [28]. By adopting $k = 0.4$ for the thesis, Ueda avoided contentious discussion about whether a new phenomenon had been observed.

Throughout the thesis, harmonic balance methods, despite their limitations, prove useful even in conditions where their justification appears weak. The fact that chaos arises every time harmonic balance fails must have given Ueda some encouragement.

An Announcement   The second selection might appear as little more than the observation that not all steady nonlinear oscillations can be explained as almost periodic oscillations. The conclusions are dressed in modesty, but the contents are rewarding: beautiful examples of tangled invariant manifold structures, including a little-known candidate for the first example of a fractal basin boundary in a differential equation. These figures are of course not mere illustrations but hard-won data. The task of constructing invariant manifolds with analog computers and electromechanical pen plotters is not an enviable one, but it would foster a solid intuitive understanding of the significance of homoclinic structure.

Randomly Transitional Phenomenon   The third selection presents Ueda's understanding of the new phenomenon now called chaotic attractors. The variety of examples underscores the widespread, generic occurrence of the phenomenon. Homoclinic structures are extensively documented, the Birkhoff-Smith theorem on periodic points is quoted, and the existence of order 2^n subharmonics is inferred via the Levinson index theorem. The term "randomly transitional phenomenon" is proposed to describe the ceaseless shifting among an infinity of unstable subharmonics.

In an Appendix, Ueda presents Birkhoff's theory of recurrence and central motions as the context in which the new steady oscillation phenomenon should be understood. Ueda was nervous about his mastery of Birkhoff's theory, but the only defect in the Appendix is omission of Pugh's results showing that generically the central motions consist solely of the nonwandering set, and thus Birkhoff's construction of higher-order nonwandering sets is unnecessary.

Ueda describes the new phenomenon as structurally stable, which would have been read as a technical error by mathematicians at that time. But surely it is the mathematical definition of structural stability which needs mending, not reality! Indeed, Zeeman has recently proposed a way forward in "Nonlinearity," Vol. 1, p. 115 (1988).

Attractor Explosion and Catastrophe   The fifth selection documents a case in which a chaotic attractor undergoes a severe bifurcation. This event later became widely known as an interior crisis due to the work of James Yorke and colleagues, who overlooked Ueda's claim to precedence.

Those who wish to reproduce Ueda's results should bear in mind the concept of structural stability and how it relates to analog and digital simulation. Ueda found this bifurcation using analog simulation, which suggested the simple picture described in Section 21.4 of R. H. Abraham and C. D. Shaw, "Dynamics: the Geometry of Behavior" (Second Edition). Upon fine checking with digital simulation, Ueda confirmed this picture, but also found that near the bifurcation threshold, the chaotic attractor collapses to periodic orbits which are stable in small windows of parameter values. Depending on the particular numerical scheme chosen, it could happen that these periodic windows might obscure the simple chaotic explosion scenario. But since the global bifurcation associated with chaotic attractor explosion lies on a generic codimension one manifold in parameter space, it would be possible to slide a small distance along this manifold to slightly different parameters where the digital simulations would reveal the same simple bifurcation suggested by the analog results.

Attractor Extinction   In this fifth selection, Ueda mentions another global bifurcation phenomenon, the "extinction of strange attractors" by a "transition chain"; this refers to his discovery of what is now termed a blue sky catastrophe or boundary crisis, first reported in [15]. That paper is not included in the present volume.

This volume concludes with the expanded 1991 version of the famous "Picture book," and a reminiscence, translated into English by Mrs. Masako Ohnuki with such skill that I had the wonderful sensation of hearing Ueda's own voice unhindered by the barrier of language.

Bruce Stewart
Upton, New York

TABLE OF CONTENTS

Selection 1

SOME PROBLEMS IN THE THEORY OF NONLINEAR OSCILLATIONS

Yoshisuke Ueda

Department of Electrical Engineering, Kyoto University

Abstract

This is the condensed version of the author's doctoral dissertation submitted to the Faculty of Engineering, Kyoto University, in February 1965. The text deals with two subjects: one is the generation of higher harmonics in simple nonlinear electrical circuits, and the other is concerned with the phenomenon of frequency entrainment which occurs in self-oscillatory systems under the impression of external periodic force.

1. HIGHER-HARMONIC OSCILLATIONS IN A SERIES-RESONANCE CIRCUIT

1.1 Introduction

Under the action of a sinusoidal external force, a nonlinear system may exhibit phenomena which are basically different from those found in linear systems. One of the salient features of such phenomena is the generation of higher harmonics and subharmonics. A considerable number of papers have been published concerning subharmonic oscillations in nonlinear systems [1-4]; however, very few investigations have been reported on the generation of higher harmonics.

This chapter deals with higher harmonic oscillations which occur in a series-resonance circuit containing a saturable inductor and a capacitor in series. The differential equation which describes the system takes the form of Duffing's equation. The amplitude characteristics of approximate periodic solutions are obtained by using the harmonic balance method, and the stability of these solutions is investigated by applying Hill's equation. These results are then tested by using analog and digital computers. An experimental result using a series-resonance circuit is cited at the end of this chapter [5]. This experiment was a motive for the present study.

1.2 Derivation of the Fundamental Equation

The schematic diagram illustrated in Fig. 1.1 shows an electrical circuit in which nonlinear oscillation takes place due to the saturable-core inductance $L(\phi)$ under the impression of the alternating voltage $E \sin \omega t$. As shown in the figure, the resistor R is paralleled with the capacitor C, so that the circuit is dissipative. With the notation of the figure, the equations for the circuit are written as

$$n\frac{d\phi}{dt} + Ri_R = E \sin \omega t$$

$$Ri_R = \frac{1}{C} \int i_C dt, \qquad i = i_R + i_C$$

(1.1)

where n is the number of turns of the inductor coil, and ϕ is the magnetic flux in the core. Then, neglecting hysteresis, we may assume the saturation curve of the form

$$i = a_1 \phi + a_3 \phi^3$$

(1.2)

where higher powers of ϕ than the third are neglected. We introduce dimensionless variables u and v, defined by

$$i = I \cdot u, \qquad \phi = \Phi \cdot v$$

(1.3)

where I and Φ are appropriate base quantities of the current and the flux, respectively. Then Eq. (1.2) becomes

$$u = \frac{a_1 \Phi}{I} v + \frac{a_3 \Phi^3}{I} v^3 = c_1 v + c_3 v^3.$$

(1.4)

Although the base quantities I and Φ can be chosen quite arbitrarily, it is preferable, for brevity of expression, to fix them by the relations

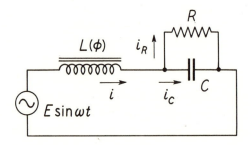

Fig. 1.1 Series-resonance circuit with nonlinear inductance.

4

Some Problems in the Theory of Nonlinear Oscillations

$$n\omega^2 C\Phi = I, \qquad c_1 + c_3 = 1. \tag{1.5}$$

Then, after elimination of i_R and i_C in Eqs. (1.1) and use of Eqs. (1.3), (1.4) and (1.5), the result in terms of v is

$$\frac{d^2v}{d\tau^2} + k\frac{dv}{d\tau} + c_1 v + c_3 v^3 = B\cos\tau \tag{1.6}$$

where

$$\tau = \omega t - \tan^{-1} k, \quad k = \frac{1}{\omega CR}, \quad B = \frac{E}{n\omega\Phi}\sqrt{1+k^2}.$$

Equation (1.6) is a well-known equation in the theory of nonlinear oscillations and is known as Duffing's equation [6].

1.3 Amplitude Characteristics of Approximate Periodic Solutions Using Harmonic Balance Method

(a) Periodic Solution Consisting of Odd-order Harmonics

As the amplitude B of the external force increases, an oscillation develops in which higher harmonics may not be ignored in comparison with the fundamental component. Since the system is symmetrical, we assume, for the time being, that these higher harmonics are of odd orders; hence a periodic solution for Eq. (1.6) may be written as

$$v_0(\tau) = x_1\sin\tau + y_1\cos\tau + x_3\sin 3\tau + y_3\cos 3\tau. \tag{1.7}$$

Terms of harmonics higher than the third are certain to be present but are ignored to this order of approximation.

The coefficients on the right side of Eq. (1.7) may be found by the method of harmonic balance [5, 7-8]; that is, substituting Eq. (1.7) into (1.6) and equating the coefficients of the terms containing $\sin\tau$, $\cos\tau$, $\sin 3\tau$ and $\cos 3\tau$ separately to zero yields

$$-A_1 x_1 - k y_1 - \frac{3}{4}c_3[(x_1^2 - y_1^2)x_3 + 2x_1 y_1 y_3] \equiv X_1(x_1, y_1, x_3, y_3) = 0$$

$$k x_1 - A_1 y_1 + \frac{3}{4}c_3[2x_1 y_1 x_3 - (x_1^2 - y_1^2)y_3] \equiv Y_1(x_1, y_1, x_3, y_3) = B$$

$$-A_3 x_3 - 3k y_3 - \frac{1}{4}c_3(x_1^2 - 3y_1^2)x_1 \equiv X_3(x_1, y_1, x_3, y_3) = 0$$

$$3k x_3 - A_3 y_3 - \frac{1}{4}c_3(3x_1^2 - y_1^2)y_1 \equiv Y_3(x_1, y_1, x_3, y_3) = 0$$

$$\tag{1.8}$$

5

where

$$A_1 = 1 - c_1 - \frac{3}{4}c_3(r_1^2 + 2r_3^2), \quad A_3 = 9 - c_1 - \frac{3}{4}c_3(2r_1^2 + r_3^2)$$

$$r_1^2 = x_1^2 + y_1^2, \qquad\qquad r_3^2 = x_3^2 + y_3^2.$$

Elimination of the x and y components in the above equations gives

$$\left[\left(A_1 - \frac{3r_3^2}{r_1^2}A_3\right)^2 + k^2\left(1 + \frac{9r_3^2}{r_1^2}\right)^2\right]r_1^2 = B^2 \tag{1.9}$$

$$(A_3^2 + 9k^2)r_3^2 = \frac{1}{16}c_3^2 r_1^6.$$

From these relations the components r_1 and r_3 of the approximate periodic solution are determined. By use of Eqs. (1.8) and (1.9) the coefficients of the periodic solution can be obtained from r_1 and r_3 by first computing

$$x_1 = \frac{k(r_1^2 + 9r_3^2)}{B}, \quad y_1 = \frac{-(A_1 r_1^2 - 3A_3 r_3^2)}{B} \tag{1.10}$$

and then subsequently

$$x_3 = \frac{4r_3^2}{c_3 r_1^6}[-PA_3 + 3kQ], \quad y_3 = \frac{4r_3^2}{c_3 r_1^6}[-QA_3 - 3kP] \tag{1.11}$$

where

$$P = (x_1^2 - 3y_1^2)x_1, \quad Q = (3x_1^2 - y_1^2)y_1.$$

(b) Stability Investigation of the Periodic Solution

The periodic states of equilibrium determined by Eqs. (1.7), (1.10) and (1.11) are not always realized, but are sustained actually if they are stable. In this section the stability of the periodic solution will be investigated by considering the behavior of a small variation $\xi(\tau)$ from the periodic solution $v_0(\tau)$. If this variation $\xi(\tau)$ tends to zero with increasing τ, the periodic solution is stable (asymptotically stable in the sense of Lyapunov [9-10]); if $\xi(\tau)$ diverges, the periodic solution is unstable. Let $\xi(\tau)$ be a small variation defined by

$$v(\tau) = v_0(\tau) + \xi(\tau). \tag{1.12}$$

Substituting Eq. (1.12) into (1.6) and neglecting terms of higher degree than the first in ξ, we obtain the variational equation

$$\frac{d^2\xi}{d\tau^2} + k\frac{d\xi}{d\tau} + (c_1 + 3c_3 v_0^2)\xi = 0. \tag{1.13}$$

6

Some Problems in the Theory of Nonlinear Oscillations

Introducing a new variable $\eta(\tau)$ defined by

$$\xi(\tau) = e^{-\delta\tau} \cdot \eta(\tau), \qquad \delta = k/2 \tag{1.14}$$

yields

$$\frac{d^2\eta}{d\tau^2} + (c_1 - \delta^2 + 3c_3 v_0^2)\eta = 0, \tag{1.15}$$

in which the first-derivative term has been eliminated. Inserting $v_0(\tau)$ as given by Eq. (1.7) into (1.15) leads to a Hill's equation of the form

$$\frac{d^2\eta}{d\tau^2} + \left(\theta_0 + 2\sum_{n=1}^{3} \theta_{ns} \sin 2n\tau + 2\sum_{n=1}^{3} \theta_{nc} \cos 2n\tau\right)\eta = 0 \tag{1.16}$$

where

$$\theta_0 = c_1 - \delta^2 + \frac{3}{2}c_3(r_1^2 + r_3^2)$$

$$\theta_{1s} = \frac{3}{2}c_3(x_1 y_1 - x_1 y_3 + y_1 x_3), \qquad \theta_{1c} = -\frac{3}{4}c_3(x_1^2 - y_1^2) + \frac{3}{2}c_3(x_1 x_3 + y_1 y_3)$$

$$\theta_{2s} = \frac{3}{2}c_3(x_1 y_3 + y_1 x_3), \qquad \theta_{2c} = -\frac{3}{2}c_3(x_1 x_3 - y_1 y_3)$$

$$\theta_{3s} = \frac{3}{2}c_3 x_3 y_3, \qquad \theta_{3c} = -\frac{3}{4}c_3(x_3^2 - y_3^2).$$

By Floquet's theorem [11], the general solution of Eq. (1.16) takes the form

$$\eta(\tau) = Ae^{\mu\tau}\phi(\tau) + Be^{-\mu\tau}\psi(\tau) \tag{1.17}$$

where A and B are arbitrary constants, $\phi(\tau)$ and $\psi(\tau)$ are periodic functions of τ of period π or 2π, and μ is the characteristic exponent to be determined by the coefficients θ_0, θ_{ns} and θ_{nc}, and may be considered to be real or imaginary, but not complex. From the theory of Hill's equation [12-14], we see that there are regions of coefficients in which the solution, Eq. (1.17), is either stable (μ: imaginary) or unstable (μ: real), and that these regions of stability and instability appear alternately as the coefficient θ_0 increases. For convenience, we shall refer to the regions of instability as the first, the second, $\cdots$ unstable regions as the coefficient θ_0 increases from zero. It is known that the periodic functions $\phi(\tau)$ and $\psi(\tau)$ in Eq. (1.17) are composed of odd-order harmonics in the regions of odd orders and even-order harmonics in the regions of even orders and that, in the n-th unstable region, the n-th harmonic component predominates over other harmonics.

7

Since Eq. (1.7) is an approximate solution of Eq. (1.6), it is appropriate to consider a solution of Eq. (1.16) approximated to the same order. Therefore we assume that a particular solution in the first and the third unstable regions is given by

$$\eta(\tau) = e^{\mu\tau}\phi(\tau) = e^{\mu\tau}[b_1 \sin(\tau - \sigma_1) + b_3 \sin(3\tau - \sigma_3)]. \qquad (1.18)$$

We substitute this into Eq. (1.16) and apply the method of harmonic balance; the resulting homogeneous system has a nontrivial solution according to Cramer's rule if

$$\Delta_1(\mu)$$

$$\equiv \begin{vmatrix} \theta_0 + \mu^2 - 1 - \theta_{1c} & \theta_{1s} - 2\mu & \theta_{1c} - \theta_{2c} & -\theta_{1s} + \theta_{2s} \\ \theta_{1s} + 2\mu & \theta_0 + \mu^2 - 1 + \theta_{1c} & \theta_{1s} + \theta_{2s} & \theta_{1c} + \theta_{2c} \\ \theta_{1c} - \theta_{2c} & \theta_{1s} + \theta_{2s} & \theta_0 + \mu^2 - 9 - \theta_{3c} & \theta_{3s} - 6\mu \\ -\theta_{1s} + \theta_{2s} & \theta_{1c} + \theta_{2c} & \theta_{3s} + 6\mu & \theta_0 + \mu^2 - 9 + \theta_{3c} \end{vmatrix}$$

$$= 0. \qquad (1.19)$$

From Eqs. (1.14) and (1.17) we see that the variation ξ tends to zero with increasing τ provided that $|\mu| < \delta$. Clearly the stability threshold occurs when $\Delta_1(\delta) = 0$. Numerical experience has established that the stability condition for the first and the third unstable regions can be taken as

$$\Delta_1(\delta) > 0. \qquad (1.20)$$

By virtue of Eqs. (1.8) and the expressions for the coefficients θ_0 and θ_{ns}, θ_{nc}, the stability condition (1.20) can be represented by

$$\Delta_1(\delta) \equiv \frac{\partial(X_1, Y_1, X_3, Y_3)}{\partial(x_1, y_1, x_3, y_3)} > 0. \qquad (1.21)$$

From this relation it is easily seen that the vertical tangency of the characteristic curves (Br_1- and Br_3-relations) occurs at the stability limit $\Delta_1(\delta) = 0$ of the first and the third unstable regions.

A particular solution of Eq. (1.16) in the second unstable region may appropriately be taken as

$$\eta(\tau) = e^{\mu\tau}\phi(\tau) = e^{\mu\tau}[b_0 + b_2 \sin(2\tau - \sigma_2)]. \qquad (1.22)$$

8

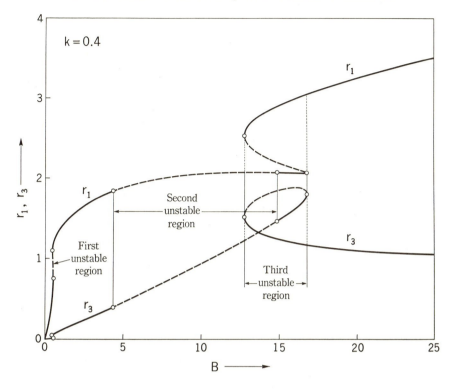

Fig. 1.2 Amplitude characteristic curves of the approximate periodic
solution given by Eq. (1.7).

Proceeding analogously as above, the characteristic exponent μ is determined
by

$$\Delta_2(\mu) \equiv \begin{vmatrix} \theta_0 + \mu^2 & \theta_{1s} & \theta_{1c} \\ 2\theta_{1s} & \theta_0 + \mu^2 - 4 - \theta_{2c} & \theta_{2s} - 4\mu \\ 2\theta_{1c} & \theta_{2s} + 4\mu & \theta_0 + \mu^2 - 4 + \theta_{2c} \end{vmatrix} = 0, \qquad (1.23)$$

and the stability condition for the second unstable region, i.e., $|\mu| < \delta$, is given
by

$$\Delta_2(\delta) > 0. \qquad (1.24)$$

Figure 1.2 shows the amplitude characteristics for solutions of the form (1.7)
calculated by numerical solution of Eqs. (1.9) with the system parameters $k =$
0.4, $c_1 = 0$ and $c_3 = 1$ in Eq. (1.6). The dashed portions of the characteristic
curves represent unstable states, and the stability condition (1.20) or (1.24) is
not satisfied in these intervals.

9

(c) Periodic Solution Including Even-order Harmonics

It has been pointed out above that, from the periodic state given by Eq. (1.7), even-order harmonics are excited in the second unstable region (see Fig. 1.2). In this region, the oscillation would gradually build up with increasing amplitude taking the form

$$e^{(\mu-\delta)\tau}\,[b_0 + b_2 \sin(2\tau - \sigma_2)] \quad \text{with} \quad \mu - \delta > 0$$

and ultimately reach the steady state with a constant amplitude which is limited by the nonlinearity of the system. This implies that, for certain intervals of B, such even-order harmonics must be considered in the periodic solution. Therefore we assume a periodic solution for Eq. (1.6) of the form

$$v_0(\tau) = z + x_1 \sin \tau + y_1 \cos \tau + x_2 \sin 2\tau + y_2 \cos 2\tau. \tag{1.25}$$

Terms of harmonics higher than the second, especially the third harmonic, are certain to be present but are ignored to avoid unwieldy calculations. The unknown coefficients on the right side of Eq. (1.25) are determined in much the same manner as before; that is, substituting Eq. (1.25) into (1.6) and equating the coefficients of the non-oscillatory term and of the terms containing $\sin \tau$, $\cos \tau$, $\sin 2\tau$ and $\cos 2\tau$ separately to zero yields

$$-A_0 z + \frac{3}{4} c_3 [2x_1 y_1 x_2 - (x_1^2 - y_1^2) y_2] \equiv Z(z, x_1, y_1, x_2, y_2) = 0$$

$$-A_1 x_1 - k y_1 + 3 c_3 z (y_1 x_2 - x_1 y_2) \equiv X_1(z, x_1, y_1, x_2, y_2) = 0$$

$$k x_1 - A_1 y_1 + 3 c_3 z (x_1 x_2 + y_1 y_2) \equiv Y_1(z, x_1, y_1, x_2, y_2) = B \tag{1.26}$$

$$-A x_2 - 2 k y_2 + 3 c_3 z x_1 y_1 \equiv X_2(z, x_1, y_1, x_2, y_2) = 0$$

$$2 k x_2 - A_2 y_2 - \frac{3}{2} c_3 z (x_1^2 - y_1^2) \equiv Y_2(z, x_1, y_1, x_2, y_2) = 0$$

where

$$A_0 = -c_1 - c_3 \left[z^2 + \frac{3}{2}(r_1^2 + r_2^2) \right]$$

$$A_1 = 1 - c_1 - \frac{3}{4} c_3 (4 z^2 + r_1^2 + 2 r_2^2), \quad A_2 = 4 - c_1 - \frac{3}{4} c_3 (4 z^2 + 2 r_1^2 + r_2^2)$$

$$r_1^2 = x_1^2 + y_1^2, \qquad\qquad r_2^2 = x_2^2 + y_2^2.$$

Elimination of the x and y components in the above equations gives

$$\left[\left(A_1 - \frac{2 r_2^2}{r_1^2} A_2 \right)^2 + k^2 \left(1 + \frac{4 r_2^2}{r_1^2} \right)^2 \right] r_1^2 = B^2$$

$$- A_0 z^2 + \frac{1}{2} A_2 r_2^2 = 0 \tag{1.27}$$

$$(A_2^2 + 4k^2) r_2^2 = \frac{9}{4} c_3^2 z^2 r_1^4.$$

From these relations z, r_1 and r_2 are determined. By use of Eqs. (1.26) and (1.27) the coefficients of the periodic solution are found, once z, r_1 and r_2 are obtained, by first computing

$$x_1 = \frac{k(r_1^2 + 4r_2^2)}{B}, \qquad y_1 = \frac{-(A_1 r_1^2 - 2A_2 r_2^2)}{B} \tag{1.28}$$

and then subsequently

$$x_2 = \frac{4r_2^2}{3c_3 z r_1^4} [A_2 x_1 y_1 + k(x_1^2 - y_1^2)]$$

$$\tag{1.29}$$

$$y_2 = \frac{4r_2^2}{3c_3 z r_1^4} [2k x_1 y_1 - \frac{1}{2} A_2 (x_1^2 - y_1^2)].$$

Proceeding analogously as before, the condition for stability may also be derived; namely, inserting $v_0(\tau)$ as given by Eq. (1.25) into (1.15) leads to a Hill's equation of the form

$$\frac{d^2\eta}{d\tau^2} + \left(\theta_0 + 2 \sum_{n=1}^{4} \theta_{ns} \sin n\tau + 2 \sum_{n=1}^{4} \theta_{nc} \cos n\tau \right) \eta = 0. \tag{1.30}$$

A particular solution of Eq. (1.30) in the second unstable region may be assumed to have the form

$$\eta(\tau) = e^{\mu\tau} \phi(\tau) = e^{\mu\tau} [b_0 + b_1 \sin(\tau - \sigma_1) + b_2 \sin(2\tau - \sigma_2)]. \tag{1.31}$$

By use of Eqs. (1.26) the stability condition is obtained as

$$\frac{\partial(Z, X_1, Y_1, X_2, Y_2)}{\partial(z, x_1, y_1, x_2, y_2)} > 0. \tag{1.32}$$

Figure 1.3 shows the amplitude characteristics of the approximate periodic solution given by Eq. (1.25). The system parameters are the same as in Fig. 1.2, i.e., $k = 0.4$, $c_1 = 0$, and $c_3 = 1$. The dashed portion in the first unstable region represents an unstable state for a solution of the form (1.25) with $z = x_2 = y_2 = 0$. The dashed portion in the second unstable region represents an unstable state for a solution (1.25) with even-order harmonics suppressed;

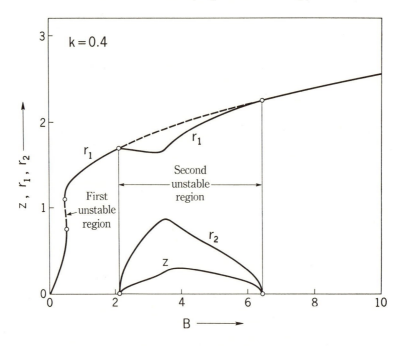

Fig. 1.3 Amplitude characteristic curves of the approximate periodic
solution given by Eq. (1.25).

indeed the term second unstable region should be understood here to refer to
an approximate solution involving only the fundamental frequency.* It is to be
noted that the second unstable region of Fig. 1.3 is narrower than that of Fig.
1.2 because the third harmonic in Eq. (1.25) was neglected. It may be expected
that if an approximate solution including constant, fundamental, second- and
third-order harmonics were computed, it would represent a steady state across
the entire second unstable region of Fig. 1.2; this is consistent with the results
of analog computer analysis presented in the next section.

It is worth mentioning that the second harmonic is sustained in the second
unstable region even though the system is symmetrical.

1.4 Analog Computer Analysis

The approximate periodic solutions obtained in the preceding section are

---

*As the coefficient of η in Eq. (1.30) contains even and odd harmonics, there are regions of
coefficients θ_0, θ_{ns} and θ_{nc} in which the $1/2, 3/2, \cdots$ harmonics are excited. This implies that
in the second unstable region of Fig. 1.3 there may exist intervals of B such that $1/2, 3/2, \cdots$
harmonics develop. A detailed investigation of such a case is, however, omitted here (cf. Secs.
1.4, 1.5 and 1.6).

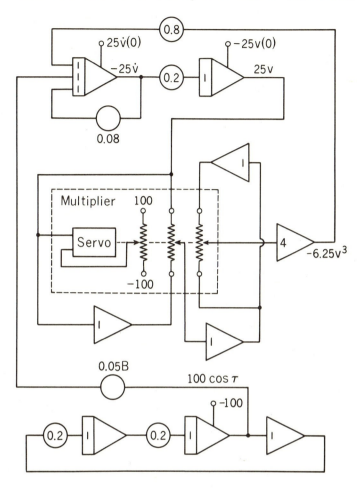

Fig. 1.4 Block diagram of an analog computer setup for the
solution of Eq. (1.33).

compared with the solutions obtained by using an analog computer. The block
diagram of Fig. 1.4 shows an analog computer setup for the solution of Eq.
(1.6), in which the system parameters k, c_1 and c_3 are set equal to the values
as given in the preceding section; i.e.,

$$\frac{d^2v}{d\tau^2} + 0.4\frac{dv}{d\tau} + v^3 = B\cos\tau. \tag{1.33}$$

The symbols in the figure follow the conventional notation; that is, the inte-
grating amplifiers in the block diagram integrate their inputs with respect to

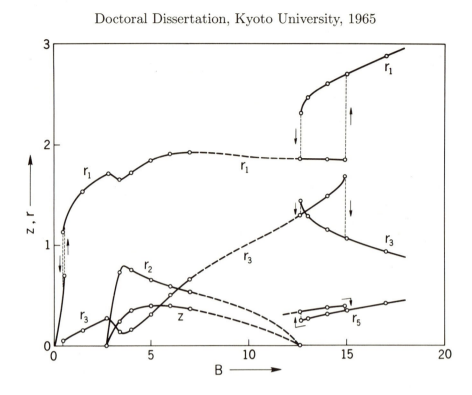

Fig. 1.5 Amplitude characteristics for the periodic solutions of Eq. (1.33) obtained by analog computer analysis.

the machine time (in seconds), which is, in this particular case, five times the independent variable τ. The solutions of Eq. (1.33) are sought for various values of B, i.e., the amplitude of the external force. From the solutions obtained in this way, each harmonic component is calculated and plotted against B in Fig. 1.5. The first unstable region ranges from $B = 0.45$ to 0.53; jump phenomena take place in the direction of arrows. The second unstable region extends from $B = 2.7$ to 12.6. In this region the occurrence of the subharmonics of order $1/2, 3/2, \cdots$ is confirmed in the interval of B approximately from 7 to 11. However, since the solutions accompanied by such subharmonics are extremely sensitive to external disturbances, the result obtained by computer analysis was not very accurate. Therefore, this region is indicated by dashed lines in Fig. 1.5. The third unstable region occurs between $B = 12.6$ to 14.9, and the oscillation jumps into another stable state on the borders of this region. These results show qualitative agreement with the results obtained in the preceding section using harmonic balance methods.

14

1.5 Digital Computer Analysis

In the preceding sections we investigated the approximate solutions of Eq. (1.6) both by using the harmonic balance method and by using an analog computer. The results thus obtained were that there are several regions of B; in the first and the third unstable regions with respect to odd harmonics there exist two stable states (see Fig. 1.2) and in the second unstable region there is only one stable state (see Fig. 1.3).[†] In this section we shall examine the periodic solutions in each unstable region by using the KDC-I digital computer.

The periodic solutions of Eq. (1.6), that is,

$$\frac{d^2v}{d\tau^2} + k\frac{dv}{d\tau} + c_1v + c_3v^3 = B\cos\tau$$

are determined by the following procedure.

The second-order differential equation (1.6) can be rewritten as simultaneous equations of the first order

$$\begin{aligned}
\frac{dv}{d\tau} &= \dot{v} \\
\frac{d\dot{v}}{d\tau} &= -k\dot{v} - c_1v - c_3v^3 + B\cos\tau.
\end{aligned} \tag{1.34}$$

We consider the location of the points whose coordinates are given by $v(\tau)$ and $\dot{v}(\tau)$ at the instants $\tau = 0, 2\pi, 4\pi, \cdots$ in the $v\dot{v}$ plane, since the right sides of Eqs. (1.34) are periodic functions in τ of period 2π. Mathematically, these points $P_n(v(2n\pi), \dot{v}(2n\pi))$ are defined as the successive images of the initial point $P_0(v(0), \dot{v}(0))$ under iterations of the mapping T from $\tau = 0$ to $2n\pi$; we denote this by [15]

$$P_n = T^n(P_0), \qquad n = 1, 2, 3, \cdots. \tag{1.35}$$

Actually, these points can be obtained approximately by performing the numerical integration of Eqs. (1.34) from $\tau = 0$ to $2n\pi$. Special attention is directed toward the location of the fixed points and the periodic points of Eq. (1.35).[‡] When an initial point P_0, the initial condition $(v(0), \dot{v}(0))$, is chosen sufficiently

---

[†]In the second unstable region, there are two oscillations differing in sign and in phase by π radians, but their amplitudes are the same.

[‡]A point whose location is invariant under the mapping is called a fixed point; i.e.,

$$P_0 = P_1 (= T(P_0))$$

and the corresponding solution $v(\tau)$ is periodic in τ with the period 2π. Periodic points are

Table 1.1 Completely stable fixed and periodic points
for Eq. (1.34) with k=0.4, c_1=0 and c_3=1

Unstable Region	B	Point	v	$\dot{v}$	h	Classification
First	0.5	1	0.2526	1.0398	$2\pi/60$	Fixed point
		2	−0.5290	0.3134	//	//
Second	4.0	1	1.5220	3.1810	$2\pi/60$	//
		2	1.8626	−1.1065	//	//
Second	9.0	1	2.9857	3.2769	$2\pi/120$	2-periodic point
		2	3.1460	2.2806	//	//
		3	2.8192	−0.7005	//	//
		4	2.9310	0.2684	//	//
Third	13.0	1	1.7823	−3.7474	$2\pi/120$	Fixed point
		2	3.5927	2.0913	//	//

near the fixed (periodic) point, the point sequence $\{P_n\}$ converges to the fixed (periodic) point as $n \to \infty$ provided the fixed (periodic) point is completely stable. In order to determine the location of a stable fixed (periodic) point, we estimate the initial condition by making use of the values obtained in the preceding sections.

Then numerical integration of Eqs. (1.34) is performed from the above initial condition until the following condition is reached:

$$| P_n - P_{n+1} | \; < \; \epsilon \quad \text{for a fixed point}$$
$$| P_n - P_{n+m} | \; < \; \epsilon \quad \text{for a periodic point} \tag{1.36}$$

where ϵ is small positive constant. In the numerical examples afterwards, $\epsilon = 10^{-5}$ is used. Because of this procedure, only the stable solutions are obtained.

Once the stable fixed (periodic) points are determined, we can easily obtain the time responses (waveforms) of $v(\tau)$ and/or $\dot{v}(\tau)$. Table 1.1 lists the values of the coordinates of the completely stable fixed and periodic points for the representative values of B given in each unstable region. The same values of the system parameters are used as in the preceding examples; i.e., $k = 0.4$,

---

defined by the following relations,

$$P_0 \neq P_i \; (1 \leq i \leq m - 1), \qquad P_0 = P_m \, (= T^m(P_0)),$$

namely, periodic points are invariant under every m-th iterate of the mapping. The corresponding solution $v(\tau)$, in this case, is also periodic in τ but its least period is equal to $2m\pi$.

16

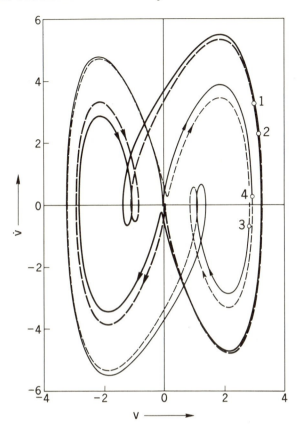

Fig. 1.6 Periodic points and associated phase plane trajectories
in the second unstable region ($B = 9.0$).

$c_1 = 0$ and $c_3 = 1$. Numerical integration of Eqs. (1.34) was carried out by using the Runge-Kutta-Gill method with the time increment h equal to $2\pi/60$ or $2\pi/120$.

There are two stable periodic solutions in each unstable region. For the first unstable region ($B = 0.5$), the fixed point 1 represents the resonant state, while the point 2 is non-resonant. For the second unstable region ($B = 4$), the fixed points 1 and 2 are symmetrically related; that is, if we indicate the periodic solution associated with the fixed point 1 by $v_{01}(\tau)$, then the fixed point 2 corresponds to the periodic solution $v_{02}(\tau) = -v_{01}(\tau - \pi)$. For the third unstable region ($B = 13$), by inspecting the magnitudes of the fundamental components of the periodic solutions represented by the fixed points 1 and 2, the point 1 corresponds to the upper branch of the r_1 characteristic curve in the third unstable region (see Figs. 1.2 and 1.5).

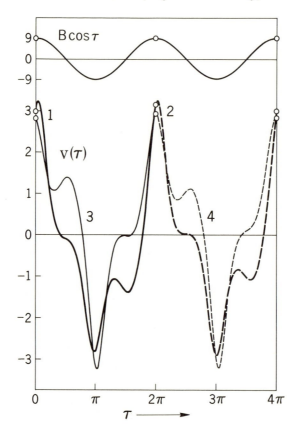

Fig. 1.7 Waveforms of the periodic solutions corresponding to Fig. 1.6.

Figure 1.6 shows the phase plane trajectories of the periodic solutions corresponding to the periodic points in the second unstable region ($B = 9$). The small circles in the figure indicate the location of the periodic points which are correlated with the subharmonic oscillations of order $1/2$. The periodic points 1 and 2 (or 3 and 4) lie on the same trajectory and, under iterations of the mapping, these points are transformed successively to the points that follow in the direction of the arrows. In order to distinguish clearly the trajectory of the point 1 to 2 (or 3 to 4) from that of the point 2 to 1 (or 4 to 3), we show the former by solid lines and the latter by dashed lines. The waveforms corresponding to the trajectories $1 \to 2 \to 1$ and $3 \to 4 \to 3$ are shown in Fig. 1.7. If we denote the periodic solution associated with the point i by $v_{0i}(\tau)$, then the periodic solutions correlated with these periodic points are given by

$$v_{01}(\tau) = v_{02}(\tau - 2\pi) = -v_{03}(\tau - \pi) = -v_{04}(\tau - 3\pi)$$

18

$$
\begin{aligned}
= -0.31 &- 0.06 \sin 1/2\tau &&+ 0.01 \cos 1/2\tau &&+ 0.58 \sin \tau &&+ 1.84 \cos \tau \\
&- 0.00 \sin 3/2\tau &&+ 0.01 \cos 3/2\tau &&+ 0.26 \sin 2\tau &&+ 0.34 \cos 2\tau \\
&+ 0.10 \sin 5/2\tau &&- 0.07 \cos 5/2\tau &&+ 0.04 \sin 3\tau &&+ 0.89 \cos 3\tau \\
&+ 0.01 \sin 7/2\tau &&+ 0.02 \cos 7/2\tau &&+ 0.11 \sin 4\tau &&+ 0.05 \cos 4\tau \\
&+ 0.03 \sin 9/2\tau &&- 0.03 \cos 9/2\tau &&+ 0.06 \sin 5\tau &&+ 0.18 \cos 5\tau \\
&+ 0.00 \sin 11/2\tau &&+ 0.00 \cos 11/2\tau &&+ 0.05 \sin 6\tau &&+ 0.02 \cos 6\tau \\
&+ 0.01 \sin 13/2\tau &&- 0.01 \cos 13/2\tau &&+ 0.02 \sin 7\tau &&+ 0.05 \cos 7\tau \\
&+ 0.00 \sin 15/2\tau &&- 0.00 \cos 15/2\tau &&+ 0.02 \sin 8\tau &&+ 0.00 \cos 8\tau \\
&+ \cdots .
\end{aligned}
$$

$$(1.37)$$

In summary, there is complete qualitative agreement between analog and digital simulations regarding the number of stable solutions, and the agreement is also quantitative to the expected accuracy.

1.6 Experimental Result

An experiment using the series-resonance circuit as illustrated in Fig. 1.1 has been performed [5, pp. 39-41]. The result was as follows.

Since B is proportional to the amplitude E of the applied voltage, varying E will bring about the excitation of higher harmonic oscillations. This is observed in Fig. 1.8, in which the effective value of the oscillating current is plotted (thick line) for a wide range of the applied voltage. By making use of a heterodyne harmonic analyzer, this current is analyzed into harmonic components. These are shown by fine lines, the numbers on which indicate the order of the harmonics. The first unstable region ranges between 24 and 40 volts of the applied voltage; the jump phenomenon in this region has been called ferro-resonance. The second unstable region extends from 180 to 580 volts. The third unstable region occurs between 660 and 670 volts, exhibiting another jump in amplitude.

Comparison of Fig. 1.8 with Fig. 1.5 shows full qualitative agreement concerning the number of stable solutions, and the amplitude variations of the harmonics up to order three. Quantitative agreement should not be expected, because of the approximation made in Eq. (1.2) assuming the saturation curve to have only linear and cubic terms.

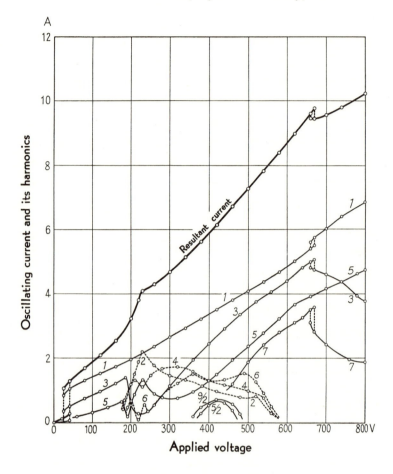

Fig. 1.8 Experimental result using a series-resonance circuit.
(Reproduced with the courtesy of Nippon Printing
and Publishing Company [5].)

2. HIGHER-HARMONIC OSCILLATIONS IN A PARALLEL-RESONANCE CIRCUIT

2.1 Introduction

In the preceding chapter, we investigated the higher-harmonic oscillations in a series-resonance circuit. Since the series condenser limits the current which magnetizes the reactor core, the applied voltage must be exceedingly high in order to bring the oscillation into the unstable regions of higher order. On the other hand, we may expect that a higher-harmonic oscillation is likely to occur in a parallel-resonance circuit because the reactor core is readily saturated under the impression of a comparatively low voltage; this will be investigated in the present chapter. The differential equation which describes the system takes the form of Mathieu's equation with additional damping and nonlinear restoring terms. An experimental result is also cited at the end of this chapter.

2.2 Derivation of the Fundamental Equation

Figure 2.1 shows the schematic diagram of a parallel-resonance circuit, in which two oscillation circuits are connected in series, each having equal values of L, R and C, respectively. With the notation of the figure, the equations for the circuit are written as

$$n\frac{d\phi_1}{dt} = Ri_{R1} = \frac{1}{C}\int i_{C1}dt$$

$$n\frac{d\phi_2}{dt} = Ri_{R2} = \frac{1}{C}\int i_{C2}dt$$

$$n\frac{d\phi_1}{dt} + n\frac{d\phi_2}{dt} = E\sin\omega t \tag{2.1}$$

$$i = i_{L1} + i_{R1} + i_{C1} = i_{L2} + i_{R2} + i_{C2}.$$

The same saturation curve is assumed for both of the inductors $L(\phi_1)$ and $L(\phi_2)$

$$i_{L1} = a_1\phi_1 + a_3\phi_1^3, \qquad i_{L2} = a_1\phi_2 + a_3\phi_2^3. \tag{2.2}$$

If the two oscillation circuits behave identically, we have, from the third of Eqs. (2.1)

$$\phi_1 = \phi_2 = -\frac{E}{2n\omega}\cos\omega t.$$

21

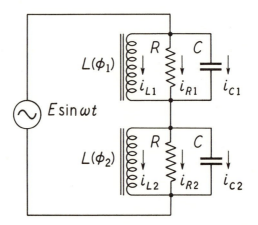

Fig. 2.1 Parallel-resonance circuit with nonlinear inductances.

An increase of the flux ϕ_1 by ϕ results in the decrease of ϕ_2 by the same amount

$$\phi_1 = -\frac{E}{2n\omega}\cos\omega t + \phi, \qquad \phi_2 = -\frac{E}{2n\omega}\cos\omega t - \phi. \tag{2.3}$$

After elimination of i_{R1}, i_{R2}, i_{C1} and i_{C2} in Eqs. (2.1) and by using Eqs. (2.3), we obtain

$$\frac{d^2\phi}{dt^2} + \frac{1}{CR}\frac{d\phi}{dt} + \frac{1}{2nC}(i_{L1} - i_{L2}) = 0. \tag{2.4}$$

Proceeding in the same manner as in Sec. 1.2, we introduce dimensionless variables defined by

$$i_{L1} = I \cdot u_{L1}, \qquad i_{L2} = I \cdot u_{L2}, \qquad \phi = \Phi \cdot v \tag{2.5}$$

and fix the base quantities I and Φ by the following relations

$$n\omega^2 C\Phi = I, \qquad c_1 + c_3 = 1 \tag{2.6}$$

where

$$c_1 = \frac{a_1\Phi}{I}, \qquad c_3 = \frac{a_3\Phi^3}{I}.$$

Then, by use of Eqs. (2.2), (2.3), (2.5) and (2.6), Eq. (2.4) may be written in normalized form as

$$\frac{d^2v}{d\tau^2} + k\frac{dv}{d\tau} + \left(c_1 + \frac{3}{2}c_3 B^2 + \frac{3}{2}c_3 B^2 \cos 2\tau\right)v + c_3 v^3 = 0 \tag{2.7}$$

Some Problems in the Theory of Nonlinear Oscillations

where
$$\tau = \omega t, \qquad k = \frac{1}{\omega C R}, \qquad B = \frac{E}{2n\omega\Phi}.$$

2.3 Amplitude Characteristics of Approximate Periodic Solutions Using Harmonic Balance Method

We assume for the moment that $k = 0$ and that v is so small that we may neglect the nonlinear term in Eq. (2.7). Then equation (2.7) reduces to the Mathieu's equation

$$\frac{d^2v}{d\tau^2} + (\theta_0 + 2\theta_1 \cos 2\tau)v = 0 \qquad (2.8)$$

where
$$\theta_0 = c_1 + \frac{3}{2}c_3 B^2, \qquad \theta_1 = \frac{3}{4}c_3 B^2.$$

From the theory of Mathieu's equation [13-14, 16] we see that there are regions of the coefficients, θ_0 and θ_1, in which the solution for Eq. (2.8) is either stable (remains bounded as τ increases) or unstable (diverges unboundedly), and that these regions of stability and instability appear alternately as the coefficient θ_0 increases. We shall call these regions of instability the first, the second, $\cdots$ unstable regions as the coefficient θ_0 increases from zero. When the coefficients θ_0 and θ_1 lie in the n-th unstable region, a higher harmonic of the n-th order is predominantly excited. Once the oscillation builds up, the nonlinear term $c_3 v^3$ in Eq. (2.7) may not be ignored. It is this term that finally prevents the amplitude of the oscillation from growing unboundedly.

(a) Periodic Solutions

After these preliminary remarks, we now proceed to investigate the periodic solution of Eq. (2.7) and assume the following form of the solution.

Harmonic: $\qquad v_0(\tau) = x_1 \sin \tau + y_1 \cos \tau$ $\qquad (2.9)$

Second-harmonic: $\quad v_0(\tau) = z + x_2 \sin 2\tau + y_2 \cos 2\tau$ $\qquad (2.10)$

Third-harmonic: $\quad v_0(\tau) = x_1 \sin \tau + y_1 \cos \tau + x_3 \sin 3\tau + y_3 \cos 3\tau$ $\quad (2.11)$

(i) Harmonic Oscillation   In order to determine the coefficients on the right side of Eq. (2.9), we use the method of harmonic balance; namely, substituting Eq. (2.9) into (2.7) and equating the coefficients of the terms containing $\sin \tau$ and $\cos \tau$ separately to zero yields

23

$$-\left(A_1 + \frac{3}{4}c_3 B^2\right)x_1 - ky_1 \equiv X_1(x_1, y_1) = 0$$

$$kx_1 - \left(A_1 - \frac{3}{4}c_3 B^2\right)y_1 \equiv Y_1(x_1, y_1) = 0$$

(2.12)

where

$$A_1 = 1 - c_1 - \frac{3}{4}c_3(2B^2 + r_1^2), \qquad r_1^2 = x_1^2 + y_1^2.$$

Elimination of the x and y components in the above equations gives

$$\left[A_1^2 + k^2 - \left(\frac{3}{4}c_3 B^2\right)^2\right]r_1^2 = 0,$$

(2.13)

from which the amplitude r_1 is found to be either

$$r_1^2 = 0$$

(2.14)

or

$$r_1^2 = \left(\frac{4}{3} - 2B^2\right) \pm \sqrt{B^4 - \left(\frac{4k}{3c_3}\right)^2}.$$

(2.15)

(ii) Second-harmonic Oscillation   Substituting Eq. (2.10) into (2.7) and equating the coefficients of the non-oscillatory term and of the terms containing $\sin 2\tau$ and $\cos 2\tau$ separately to zero, we obtain

$$-A_0 z + \frac{3}{4}c_3 B^2 y_2 \equiv Z(z, x_2, y_2) = 0$$

$$- A_2 x_2 - 2ky_2 \equiv X_2(z, x_2, y_2) = 0$$

(2.16)

$$2kx_2 - A_2 y_2 + \frac{3}{2}c_3 B^2 z \equiv Y_2(z, x_2, y_2) = 0$$

where

$$A_0 = -c_1 - c_3[z^2 + \frac{3}{2}(B^2 + r_2^2)], \quad A_2 = 4 - c_1 - \frac{3}{4}c_3(2B^2 + 4z^2 + r_2^2)$$

$$r_2^2 = x_2^2 + y_2^2.$$

Elimination of the x and y components in the above equations gives

$$-A_0 z^2 + \frac{1}{2}A_2 r_2^2 = 0$$

$$(A_2^2 + 4k^2)r_2^2 = \left(\frac{3}{2}c_3 B^2\right)^2 z^2$$

(2.17)

from which the unknown quantities z and r_2 are determined.

(iii) Third-harmonic Oscillation   Substituting Eq. (2.11) into (2.7) and equating the terms containing $\sin \tau$, $\cos \tau$, $\sin 3\tau$ and $\cos 3\tau$ separately to zero, we obtain

$$-\left(A_1 + \frac{3}{4}c_3 B^2\right)x_1 - ky_1 - \frac{3}{4}c_3[(x_1^2 - y_1^2 - B^2)x_3 + 2x_1 y_1 y_3]$$

$$\equiv X_1(x_1, y_1, x_3, y_3) = 0$$

$$kx_1 - \left(A_1 - \frac{3}{4}c_3 B^2\right)y_1 + \frac{3}{4}c_3[2x_1 y_1 x_3 - (x_1^2 - y_1^2 - B^2)y_3]$$

$$\equiv Y_1(x_1, y_1, x_3, y_3) = 0 \quad (2.18)$$

$$-A_3 x_3 - 3ky_3 + \frac{1}{4}c_3[3B^2 - (x_1^2 - 3y_1^2)]x_1 \equiv X_3(x_1, y_1, x_3, y_3) = 0$$

$$3kx_3 - A_3 y_3 + \frac{1}{4}c_3[3B^2 - (3x_1^2 - y_1^2)]y_1 \equiv Y_3(x_1, y_1, x_3, y_3) = 0$$

where

$$A_1 = 1 - c_1 - \frac{3}{4}c_3(2B^2 + r_1^2 + 2r_3^2), \quad A_3 = 9 - c_1 - \frac{3}{4}c_3(2B^2 + 2r_1^2 + r_3^2)$$

$$r_1^2 = x_1^2 + y_1^2, \qquad\qquad\qquad r_3^2 = x_3^2 + y_3^2$$

from which the unknown quantities x_1, y_1, x_3 and y_3, and consequently the amplitudes, r_1 and r_3, are determined.

(b) Stability Investigation of the Periodic Solutions

The periodic solutions given above are sustained actually only when they are stable. Here the stability of the periodic solutions will be investigated in the same manner as we have done in Sec. 1.3. We consider a small variation $\xi(\tau)$ from the periodic solution $v_0(\tau)$. Then the behavior of $\xi(\tau)$ is described by the following variational equation

$$\frac{d^2\xi}{d\tau^2} + k\frac{d\xi}{d\tau} + \left(c_1 + \frac{3}{2}c_3 B^2 + \frac{3}{2}c_3 B^2 \cos 2\tau + 3c_3 v_0^2\right)\xi = 0. \quad (2.19)$$

Furthermore we introduce a new variable $\eta(\tau)$ defined by

$$\xi(\tau) = e^{-\delta\tau} \cdot \eta(\tau), \qquad \delta = k/2 \quad (2.20)$$

to remove the first-derivative term. Then we obtain

$$\frac{d^2\eta}{d\tau^2} + \left(c_1 - \delta^2 + \frac{3}{2}c_3 B^2 + \frac{3}{2}c_3 B^2 \cos 2\tau + 3c_3 v_0^2\right)\eta = 0. \quad (2.21)$$

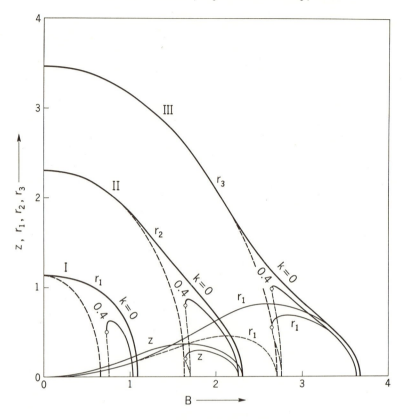

Fig. 2.2 Amplitude characteristic curves of the approximate periodic
solutions given by Eqs. (2.9), (2.10) and (2.11).

(i) Stability Condition for the Harmonic Oscillation   Inserting $v_0(\tau)$ as
given by Eq. (2.9) into (2.21) leads to

$$\frac{d^2\eta}{d\tau^2} + (\theta_0 + 2\theta_{1s}\sin 2\tau + 2\theta_{1c}\cos 2\tau)\eta = 0 \qquad (2.22)$$

where

$$\theta_0 = c_1 - \delta^2 + \frac{3}{2}c_3(B^2 + r_1^2)$$

$$\theta_{1s} = \frac{3}{2}c_3 x_1 y_1, \qquad \theta_{1c} = \frac{3}{4}c_3[B^2 - (x_1^2 - y_1^2)].$$

We assume that a particular solution of Eq. (2.22) in the first unstable region
is given by

$$\eta(\tau) = e^{\mu\tau}\phi(\tau) = e^{\mu\tau}\sin(\tau - \sigma_1). \qquad (2.23)$$

26

Some Problems in the Theory of Nonlinear Oscillations

Proceeding analogously as in Sec. 1.3, the stability condition $|\mu| < \delta$ leads to

$$\Delta_1(\delta) \equiv \begin{vmatrix} \theta_0 + \delta^2 - 1 - \theta_{1c} & \theta_{1s} - 2\delta \\ \theta_{1s} + 2\delta & \theta_0 + \delta^2 - 1 + \theta_{1c} \end{vmatrix} \equiv \frac{\partial(X_1, Y_1)}{\partial(x_1, y_1)} > 0. \qquad (2.24)$$

(ii) Stability Conditions for the Higher-harmonic Oscillations   The conditions for stability of the solutions given by Eqs. (2.10) and Eq. (2.11) may also be derived by the same procedure as above. The results are as follows.
Stability condition for the solution (2.10):

$$\Delta_2(\delta) \equiv \frac{\partial(Z, X_2, Y_2)}{\partial(z, x_2, y_2)} > 0. \qquad (2.25)$$

Stability condition for the solution (2.11):

$$\Delta_3(\delta) \equiv \frac{\partial(X_1, Y_1, X_3, Y_3)}{\partial(x_1, y_1, x_3, y_3)} > 0. \qquad (2.26)$$

The vertical tangency of the characteristic curves (Bz-, Br_1-, Br_2- and Br_3-relations) also occurs at the stability limit $\Delta_n(\delta) = 0$ ($n = 1, 2, 3$).

Figure 2.2 shows amplitude characteristics of Eqs. (2.9), (2.10) and (2.11) for the choices of system parameters $k = 0$ and 0.4, $c_1 = 0$ and $c_3 = 1$ in Eq. (2.7). The dashed portions of the characteristic curves represent unstable states. The case $k = 0$ corresponds to the pairs of curves extending in thin tongues to the vertical axis at $B = 0$, while the case $k = 0.4$ has the stable and unstable branches joining smoothly in vertical tangency at the small circles. It is to be mentioned that the portions of the B axis interposed between the end points of the paired characteristic curves are unstable. We see in the figure that

Table 2.1 Completely stable fixed and periodic points
for Eq. (2.7) with $k = 0.4$, $c_1 = 0$ and $c_3 = 1$

Unstable Region	B	Point	v	$\dot{v}$	h	Classification*
First	0.8	1	0.2925	-0.6621	$\pi/30$	2-periodic point
		2	-0.2925	0.6621	//	//
Second	1.8	1	0.4430	0.6727	$\pi/30$	Fixed point
		2	-0.4430	-0.6727	//	//
Third	2.8	1	0.1495	-1.7041	$\pi/60$	2-periodic point
		2	-0.1495	1.7041	//	//

*The mapping T is defined at the instants $\tau = 0, \pi, 2\pi, \cdots$.

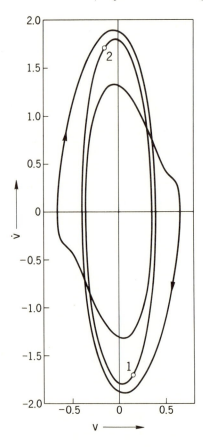

Fig. 2.3 Periodic points and associated phase plane
trajectory in the third unstable region ($B = 2.8$).

increasing B will bring about the excitation of higher-harmonic oscillations and
that once the oscillation is started, it may only be stopped by decreasing B to a
value which is lower than before, thus exhibiting the phenomenon of hysteresis.

2.4 Analog and Digital Computer Analyses

The approximate periodic solutions obtained in the preceding section have
been compared with solutions obtained by using analog and digital computers.
The methods of analysis are the same as in Secs. 1.4 and 1.5, and results show
good agreement with the results given in Fig. 2.2. Table 2.1 lists the values
of the coordinates of the completely stable fixed and periodic points for representative values of B in each unstable region. The same values of the system
parameters are used as in Fig. 2.2; i.e., $k = 0.4$, $c_1 = 0$ and $c_3 = 1$. Figure 2.3

28

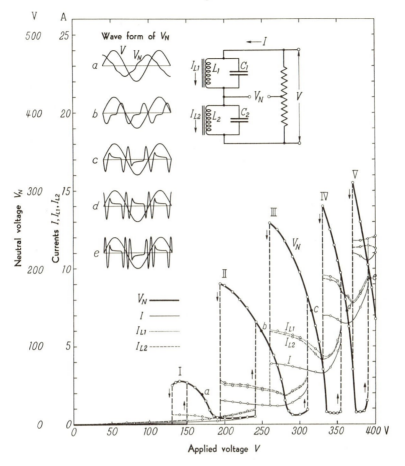

Fig. 2.4 Experimental result using a parallel-resonance circuit.
(Reproduced with the courtesy of Nippon Printing
and Publishing Company [5].)

shows the phase plane trajectory of the periodic solution corresponding to the
periodic points for the third unstable region ($B = 2.8$).

2.5 Experimental Result

An experiment on the circuit of Fig. 2.1 has been performed [5, pp. 44-48].
The result was as follows.

The excitation of the fundamental and higher-harmonic oscillations was
observed under varying E. As a result of this excitation, the potential of the
junction point of the two resonance circuits oscillates with respect to the neutral
point of the applied voltage with the frequency of the harmonic. In Fig. 2.4, the

anomalous neutral voltage V_N (which is related to the flux ϕ) is shown against the applied voltage.*

---

*The excited oscillation in the first unstable region (marked by I) has the same frequency as that of the applied voltage. See the waveform (a) in the figure. This phenomenon is known as neutral inversion in electric transmission lines.

3. PHENOMENON OF FREQUENCY ENTRAINMENT

3.1 Introduction

When a periodic force is applied to a self-oscillatory system, the frequency of the self-excited oscillation, that is, the natural frequency of the system, falls in synchronism with the driving frequency, provided these two frequencies are not far different [17-21]. This phenomenon of frequency entrainment may also occur when the ratio of the two frequencies in either order is in the neighborhood of an integer (different from unity) [22]. Thus, if the amplitude and frequency of the external force are appropriately chosen, the natural frequency of the system is entrained by a frequency which is an integral multiple or submultiple of the driving frequency. If the ratio of these two frequencies is not in the neighborhood of an integer, one may expect the occurrence of an almost periodic oscillation [23-24]. It is a salient feature of an almost periodic oscillation that the amplitude and phase of the oscillation vary slowly but periodically even in the steady state. However, the period of the amplitude variation is not an integral multiple of the period of the external force; the ratio of these two periods is in general incommensurable. Therefore, as a whole, there is no periodicity in the almost periodic oscillation.

In this chapter, response curves of the harmonic entrainment, the regions of entrainment at various frequencies, and almost periodic oscillations are studied by using the averaging method. These results are checked by using analog and digital computers.

3.2 Van der Pol's Equation with Forcing Term

In the preceding chapters we treated the cases in which the restoring force of the system was nonlinear. In this chapter we consider a system in which the nonlinearity appears in the damping. The system considered is governed by

$$\frac{d^2u}{dt^2} - \epsilon(1 - u^2)\frac{du}{dt} + u = B\cos\nu t + B_0 \tag{3.1}$$

where ϵ is a small positive constant and the right side represents an external force containing a non-oscillatory component. The left side of this equation takes the form of van der Pol's equation [25-26]. Introduction of a new variable defined by $v = u - B_0$ yields an alternative form of Eq. (3.1)

$$\frac{d^2v}{dt^2} - \mu(1 - \beta v - \gamma v^2)\frac{dv}{dt} + v = B\cos\nu t \tag{3.2}$$

where

$$\mu = (1 - B_0^2)\epsilon, \quad \beta = \frac{2B_0}{1 - B_0^2}, \quad \gamma = \frac{1}{1 - B_0^2}.$$

Here let us introduce approximate expressions for the entrained oscillations. Since μ is small, we see that when $B = 0$ the natural frequency of the system (3.2) is nearly equal to unity. Hence, when the driving frequency ν is in the neighborhood of unity, we may expect an entrained oscillation at the driving frequency ν, that is, an occurrence of harmonic entrainment. The entrained harmonic oscillation $v_0(t)$ may be expressed approximately by

$$v_0(t) = b_1 \sin \nu t + b_2 \cos \nu t. \tag{3.3}$$

On the other hand, when the driving frequency ν is far different from unity, we may expect an occurrence of higher-harmonic or subharmonic entrainment. In this case, the entrained oscillation has a frequency which is an integral multiple or a submultiple of the driving frequency ν. An approximate solution for Eq. (3.2) may be expressed by

$$v_0(t) = \frac{B}{1 - \nu^2} \cos \nu t + b_1 \sin n\nu t + b_2 \cos n\nu t \tag{3.4}$$

where

$$n = 2 \text{ or } 3: \qquad \text{for higher-harmonic oscillations,}$$
$$n = 1/2 \text{ or } 1/3: \quad \text{for subharmonic oscillations.}$$

The first term on the right side represents the forced oscillation at the driving frequency ν. The second and the third terms represent the entrained oscillation at the frequency $n\nu$, which is close to unity.

3.3 Analysis of the Equation Using the Averaging Method

(a) Derivation of an Autonomous System [10, 20, 27-29]

We now write the differential equation (3.2) in a simultaneous form

$$\frac{dv}{dt} = \dot{v}$$
$$\frac{d\dot{v}}{dt} = \mu(1 - \beta v - \gamma v^2)\dot{v} - v + B \cos \nu t. \tag{3.5}$$

The behavior of the system is described by the movement of a representative point $(v(t), \dot{v}(t))$ along the solution curves of Eqs. (3.5) in the $v\dot{v}$ plane. These solution curves are called trajectories of the representative point. Let us first consider the case in which the driving frequency ν of the external force is in the neighborhood of unity.* According to the form of the solution (3.3) considered in Sec. 3.2, we introduce a new coordinate system $(b_1(t), b_2(t))$ defined by

$$v(t) = b_1(t)\sin\nu t + b_2(t)\cos\nu t$$
$$\dot{v}(t) = \nu b_1(t)\cos\nu t - \nu b_2(t)\sin\nu t \qquad (3.6)$$

which rotates with angular frequency ν. It may therefore be conjectured that the coordinates $(b_1(t), b_2(t))$ of the representative point vary rather slowly in comparison with $(v(t), \dot{v}(t))$. To see this let us transform Eqs. (3.5) by using Eqs. (3.6). Hence

$$\frac{dx_1}{dt} = \frac{\mu}{2}\left\{\left[(1 - r_1^2)x_1 - \sigma_1 y_1 + \frac{B}{\mu\nu a_0}\right]\right.$$

$$- \frac{1}{2}\beta a_0(x_1^2 - y_1^2)\sin\nu t - \beta a_0 x_1 y_1 \cos\nu t$$

$$+ [-\sigma_1 x_1 - (1 + 2x_1^2 - 2y_1^2)y_1]\sin 2\nu t$$

$$+ \left[(1 - 4y_1^2)x_1 - \sigma_1 y_1 + \frac{B}{\mu\nu a_0}\right]\cos 2\nu t$$

$$- \frac{1}{2}\beta a_0(x_1^2 - y_1^2)\sin 3\nu t - \beta a_0 x_1 y_1 \cos 3\nu t$$

$$\left. - (3x_1^2 - y_1^2)y_1 \sin 4\nu t + (x_1^2 - 3y_1^2)x_1 \cos 4\nu t\right\}$$

$$\qquad (3.7)$$

$$\frac{dy_1}{dt} = \frac{\mu}{2}\left\{[\sigma_1 x_1 + (1 - r_1^2)y_1]\right.$$

$$- \beta a_0 x_1 y_1 \sin\nu t + \frac{1}{2}\beta a_0(x_1^2 - y_1^2)\cos\nu t$$

$$+ \left[-(1 - 2x_1^2 + 2y_1^2)x_1 + \sigma_1 y_1 - \frac{B}{\mu\nu a_0}\right]\sin 2\nu t$$

$$+ [-\sigma_1 x_1 - (1 - 4x_1^2)y_1]\cos 2\nu t$$

$$+ \beta a_0 x_1 y_1 \sin 3\nu t - \frac{1}{2}\beta a_0(x_1^2 - y_1^2)\cos 3\nu t$$

$$\left. - (x_1^2 - 3y_1^2)x_1 \sin 4\nu t - (3x_1^2 - y_1^2)y_1 \cos 4\nu t\right\}$$

---

*It is here assumed that $\nu - 1 = O(\mu)$ and $B = O(\mu)$.

where

$$x_1 = \frac{b_1}{a_0}, \quad y_1 = \frac{b_2}{a_0}, \quad r_1^2 = x_1^2 + y_1^2$$

$$a_0 = \sqrt{\frac{4}{\gamma}}, \quad \sigma_1 = \frac{1-\nu^2}{\mu\nu} \ (\text{detuning}).$$

From the form of the right sides of Eqs. (3.7), it is seen that both dx_1/dt and dy_1/dt are proportional to the small parameter μ, so that x_1 and y_1 will be slowly varying functions of t as we expected. Moreover dx_1/dt and dy_1/dt are periodic functions of t with period $2\pi/\nu$. It may therefore be considered that $x_1(t)$ and $y_1(t)$ remain approximately constant during one period $2\pi/\nu$. Hence averaging the right sides of Eqs. (3.7) over the period $2\pi/\nu$, we obtain the relations to determine dx_1/dt and dy_1/dt to a first approximation as

$$
\begin{aligned}
\frac{dx_1}{dt} &= \frac{\mu}{2}\left[(1 - r_1^2)x_1 - \sigma_1 y_1 + \frac{B}{\mu\nu a_0}\right] \equiv X_1(x_1,\, y_1) \\
\frac{dy_1}{dt} &= \frac{\mu}{2}\left[\sigma_1 x_1 + (1 - r_1^2)y_1\right] \qquad\qquad \equiv Y_1(x_1,\, y_1).
\end{aligned}
\tag{3.8}
$$

Equations (3.8) play an important role in the present investigation, since the singular points of this system correspond to the entrained harmonic oscillations and the limit cycles, if they exist, correspond to the asynchronized almost periodic oscillations. It is to be noted that x_1 and y_1 in Eqs. (3.8) denote the normalized amplitudes of the entrained oscillation since the constant a_0 represents the amplitude of the self-excited oscillation to a first approximation.

By the same procedure as above, we proceed next to derive an autonomous system for the cases in which the frequency ν of the external force or its inverse is in the neighborhood of an integer (different from unity). In this case we make use of the transformation defined by

$$
\begin{aligned}
v(t) &= \frac{B}{1-\nu^2}\cos\nu t + b_1(t)\sin n\nu t + b_2(t)\cos n\nu t \\
\dot{v}(t) &= \frac{-\nu B}{1-\nu^2}\sin\nu t + n\nu b_1(t)\cos n\nu t - n\nu b_2(t)\sin n\nu t.
\end{aligned}
\tag{3.9}
$$

Then the derived autonomous systems are as follows.

$n = 2$:

$$
\begin{aligned}
\frac{dx_2}{dt} &= \frac{\mu}{2}\left[(D - r_2^2)x_2 - \sigma_2 y_2\right] \qquad\qquad \equiv X_2(x_2,\, y_2) \\
\frac{dy_2}{dt} &= \frac{\mu}{2}\left[\sigma_2 x_2 + (D - r_2^2)y_2 - \frac{\beta}{4a_0}A^2\right] \equiv Y_2(x_2,\, y_2)
\end{aligned}
\tag{3.10}
$$

$n = 3$:

$$\frac{dx_3}{dt} = \frac{\mu}{2} [(D - r_3^2)x_3 - \sigma_3 y_3] \qquad\qquad \equiv X_3(x_3, y_3)$$

$$\frac{dy_3}{dt} = \frac{\mu}{2} \left[\sigma_3 x_3 + (D - r_3^2)y_3 - \frac{\gamma}{12a_0}A^3 \right] \equiv Y_3(x_3, y_3)$$

$$(3.11)$$

$n = 1/2$:

$$\frac{dx_{1/2}}{dt} = \frac{\mu}{2} \left[\left(D - r_{1/2}^2 + \frac{1}{2}\beta A \right)x_{1/2} - \sigma_{1/2}y_{1/2} \right] \equiv X_{1/2}(x_{1/2}, y_{1/2})$$

$$\frac{dy_{1/2}}{dt} = \frac{\mu}{2} \left[\sigma_{1/2}x_{1/2} + \left(D - r_{1/2}^2 - \frac{1}{2}\beta A \right)y_{1/2} \right] \equiv Y_{1/2}(x_{1/2}, y_{1/2})$$

$$(3.12)$$

$n = 1/3$:

$$\frac{dx_{1/3}}{dt} = \frac{\mu}{2} \left[(D - r_{1/3}^2)x_{1/3} - \sigma_{1/3}y_{1/3} + 2\frac{A}{a_0}x_{1/3}y_{1/3} \right] \qquad \equiv X_{1/3}(x_{1/3}, y_{1/3})$$

$$\frac{dy_{1/3}}{dt} = \frac{\mu}{2} \left[\sigma_{1/3}x_{1/3} + (D - r_{1/3}^2)y_{1/3} + \frac{A}{a_0}(x_{1/3}^2 - y_{1/3}^2) \right] \equiv Y_{1/3}(x_{1/3}, y_{1/3})$$

$$(3.13)$$

where

$$x_n = \frac{b_1}{a_0}, \quad y_n = \frac{b_2}{a_0}, \quad r_n^2 = x_n^2 + y_n^2, \quad a_0 = \sqrt{\frac{4}{\gamma}}$$

$$A = \frac{B}{1 - \nu^2}, \quad D = 1 - \frac{2A^2}{a_0^2}, \quad \sigma_n = \frac{1 - (n\nu)^2}{\mu n\nu} \text{ (detuning)}.$$

(b) Singular Points Correlated with Periodic Oscillations

Let x_{10} and y_{10} be the coordinates of the singular point of Eqs. (3.8). They are obtained by putting $dx_1/dt = 0$ and $dy_1/dt = 0$; i.e.,

$$X_1(x_{10}, y_{10}) = 0, \qquad Y_1(x_{10}, y_{10}) = 0 \qquad\qquad (3.14)$$

and represent the particular solutions corresponding to the equilibrium states of this system. The variational equation for these solutions is of the form

$$\frac{d\xi}{dt} = a_1\xi + a_2\eta, \qquad \frac{d\eta}{dt} = b_1\xi + b_2\eta \qquad\qquad (3.15)$$

with $a_1 = (\partial X_1/\partial x_1)_0$, $a_2 = (\partial X_1/\partial y_1)_0$, $b_1 = (\partial Y_1/\partial x_1)_0$, and $b_2 = (\partial Y_1/\partial y_1)_0$, where $(\partial X_1/\partial x_1)_0, \cdots, (\partial Y_1/\partial y_1)_0$ denote the values of $\partial X_1/\partial x_1, \cdots, \partial Y_1/\partial y_1$ at $x_1 = x_{10}$ and $y_1 = y_{10}$, respectively, and are constants.

35

Let us assume that the characteristic equation of this system has no root the real part of which is equal to zero. It is known that in this case the system (3.7) has for sufficiently small μ one and only one periodic solution which reduces to the solution $x_1 = x_{10}$ and $y_1 = y_{10}$ for $\mu = 0$. Moreover the stability of this solution is decided by the signs of the real parts of the corresponding characteristic roots. That is, if the real parts of the roots of the characteristic equation of the system (3.15) are negative, the corresponding periodic solution is stable; if at least one root has a positive real part, the periodic solution is unstable [27].

The coordinates of the singular point are given by

$$x_{10} = -\frac{\mu \nu a_0}{B}(1 - r_{10}^2)r_{10}^2, \qquad y_{10} = \frac{\mu \nu a_0}{B}\sigma_1 r_{10}^2 \qquad (3.16)$$

where r_{10}^2 is determined by the equation

$$[(1 - r_{10}^2)^2 + \sigma_1^2]r_{10}^2 = \left(\frac{B}{\mu \nu a_0}\right)^2. \qquad (3.17)$$

Equation (3.17) yields what we call the amplitude characteristics (response curves) for the harmonic oscillation and is obtained by eliminating x_{10} and y_{10} from Eqs. (3.14). Figure 3.1 is obtained by plotting Eq. (3.17) in the $\sigma_1 r_{10}^2$ plane for various values of the magnitude $(B/\mu \nu a_0)^2$. Evidently the curves are symmetrical with respect to the r_{10}^2 axis. Each point on these curves yields the amplitude r_{10}, which is correlated with the frequency ν of a possible harmonic oscillation for a given value of the amplitude B.

The singular points and the relations representing the amplitude characteristics of the entrained oscillations for the derived autonomous systems (3.10), (3.11), (3.12) and (3.13) are easily obtained in the same manner as above.

(c) Conditions for Stability of Singular Points

The periodic states of equilibrium of the initial system (3.2) are not always realized, but are actually able to exist only so long as they are stable. We have already seen that the stability of the harmonic solution of Eq. (3.2) is to be decided in accordance with the characteristic roots of the corresponding singular point (3.16). Here we will therefore consider the stability condition for the singular point.

Let ξ and η be small variations from the singular point defined by

$$x_1 = x_{10} + \xi, \qquad y_1 = y_{10} + \eta \qquad (3.18)$$

and let us determine whether these variations approach zero or not with the increase of time t. We again write the variational equations (3.15) which are

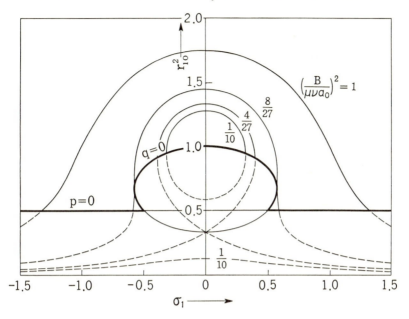

Fig. 3.1 Response curves for the harmonic oscillation.

obtained by substituting Eqs. (3.18) into (3.8) and neglecting terms of higher order than the first in ξ and η

$$\frac{d\xi}{dt} = a_1\xi + a_2\eta, \qquad \frac{d\eta}{dt} = b_1\xi + b_2\eta \qquad (3.19)$$

where

$$a_1 = \frac{\mu}{2}(1 - r_{10}^2 - 2x_{10}^2), \qquad a_2 = \frac{\mu}{2}(-\sigma_1 - 2x_{10}y_{10})$$

$$b_1 = \frac{\mu}{2}(\sigma_1 - 2x_{10}y_{10}), \qquad b_2 = \frac{\mu}{2}(1 - r_{10}^2 - 2y_{10}^2).$$

The characteristic equation of the system (3.19) is given by

$$\begin{vmatrix} a_1 - \lambda & a_2 \\ b_1 & b_2 - \lambda \end{vmatrix} = \lambda^2 + p\lambda + q = 0 \qquad (3.20)$$

where

$$p = -(a_1 + b_2) = \mu(2r_{10}^2 - 1)$$

$$q = a_1b_2 - a_2b_1 = \frac{\mu^2}{4}\left[(1 - r_{10}^2)(1 - 3r_{10}^2) + \sigma_1^2\right].$$

The variations ξ and η approach zero with the time t, provided that the real parts of λ are negative. This stability condition is given by the Routh-Hurwitz criterion [30], that is

$$p > 0 \qquad \text{and} \qquad q > 0. \tag{3.21}$$

On Fig. 3.1 the stability limits $p = 0$ and $q = 0$ are also drawn; the curve $p = 0$ is a horizontal line $r_{10}^2 = 1/2$, and the curve $q = 0$ is an ellipse which is the locus of the vertical tangents of the response curves. The unstable portions of the response curves are shown dashed in the figure. We can obtain the region of harmonic entrainment in the $B\nu$ plane by referring to the stability limit (drawn as a thick line in the figure) of Fig. 3.1. The portion of the ellipse $q = 0$ applies in the case where the amplitude B and consequently the detuning σ_1 are comparatively small, while if B and σ_1 are large the stability limit $p = 0$ applies. For intermediate values of B and σ_1 some complicated phenomena may occur, but we will not enter this problem here. A detailed investigation of such cases is reported by Cartwright [18]. See also [19].

The stability conditions for the singular points of the autonomous systems (3.10) to (3.13) are derived in the same manner as above.

(d) Regions of Frequency Entrainment

Thus far, the singular points of the derived autonomous systems (3.8), (3.10), (3.11), (3.12) and (3.13) and the relations representing the amplitude characteristic of the entrained oscillations have been investigated. The stability for these singular points has also been investigated by making use of the Routh-Hurwitz criterion. From these results we can obtain the regions of frequency entrainment on the $B\nu$ plane; namely, if the amplitude B and the frequency ν of the external force are given in these regions, the corresponding autonomous system possesses at least one stable singularity. Consequently, entrainment occurs at the corresponding harmonic, higher-harmonic, or subharmonic frequency of the external force. Figure 3.2 shows an example of the regions of frequency entrainment, computed by numerical evaluation of the stability conditions (3.21) for the derived autonomous system (3.8) and the analogous conditions for (3.10), (3.11), (3.12) and (3.13). The system parameters under consideration are

$$\epsilon = 0.2 \qquad \text{and} \qquad B_0 = 0.5$$

as in Eq. (3.1). Consequently, the parameters in Eq. (3.2) would be

$$\mu = 0.15, \qquad \beta = 4/3 \qquad \text{and} \qquad \gamma = 4/3.$$

We see that higher-harmonic or subharmonic entrainment occurs within a narrow range of the driving frequency ν. On the other hand, harmonic entrainment

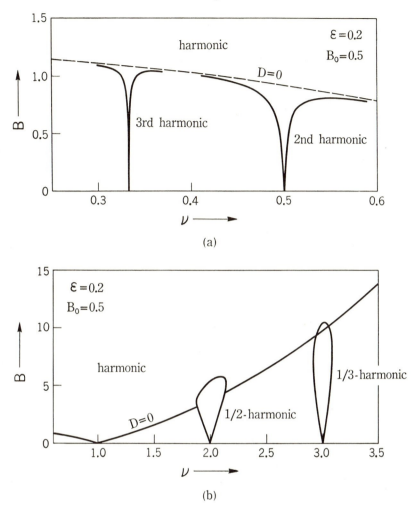

Fig. 3.2 Regions of frequency entrainment.
(a) Harmonic and higher-harmonic entrainments.
(b) Harmonic and subharmonic entrainments.

occurs at any driving frequency ν provided the amplitude B of the external force is sufficiently large.

In Fig. 3.2(a) the boundary curves of the higher-harmonic entrainment tend asymptotically to the curve $D = 0$ as the detuning σ_n $(n = 2, 3)$ increases. In the figure the curve $D = 0$ is plotted as a dashed line. We see that the inequality

$$D = 1 - \frac{2A^2}{a_0^2} < 0 \tag{3.22}$$

is equivalent to the first condition of (3.21) which gives the boundary of harmonic entrainment for large detuning σ_1.[†] It is easily ascertained that if $D < 0$, the higher-harmonic oscillations are stable. Furthermore since there are no abrupt changes in the amplitudes of the higher-harmonic components of an oscillation at the curve $D = 0$, the boundary curve $D = 0$ of the harmonic entrainment has practically no significance for the higher-harmonic entrainment.

In Fig. 3.2(b), the boundary of harmonic entrainment in the neighborhood of $\nu = 1$ $(\sigma_1 = 0)$ is given by the second condition of (3.21) for small detuning σ_1 and by the first condition for large detuning σ_1.[‡]

It is also mentioned that the regions of harmonic and 1/3-harmonic entrainments have an overlapping area. In this area common to the two regions, both the harmonic and the 1/3-harmonic oscillations are sustained. On the other hand, such a situation does not occur for the 1/2-harmonic entrainment.

(e) Limit Cycles Correlated with Almost Periodic Oscillations

The oscillations governed by van der Pol's equation with forcing term are characterized by the behavior of the representative point of the derived autonomous systems within the accuracy of the approximation made in averaging. Now suppose that we fix a point $(x_n(0), y_n(0))$ in the $x_n y_n$ plane as an initial condition. Then the representative point moves, with the increase of time t, along the integral curve which emanates from the initial point and may lead ultimately to a stable singular point. Thus the transient solutions are correlated with the integral curves, and the stationary periodic solutions, with the singular points in the $x_n y_n$ plane. However, the representative point may not always lead to a singular point, but may tend to a closed trajectory along which it moves permanently. An isolated closed trajectory such that no trajectory sufficiently near it is also closed is called a limit cycle [31]. In such a case we see that $x_n(t)$ and $y_n(t)$ tend to periodic functions having the same period in t and hence the solution of the original differential equation (3.2) will be one in which the amplitude and the phase after the lapse of sufficient time

---

[†]Since μ is small, the condition $\sigma_1^2 \gg (1-r_{10}^2)^2$ is satisfied when ν is not in the neighborhood of unity. Under this condition, we obtain from Eq. (3.17) the following approximation:

$$a_0^2 r_{10}^2 \simeq \left(\frac{B}{1-\nu^2}\right)^2 = A^2.$$

With the above result the first condition of (3.21) becomes the inequality (3.22). It is to be noted that the assumptions made in the derivation of Eqs. (3.8) become inappropriate as the detuning σ_1 increases.

[‡]As we already noted, for intermediate values of σ_1, the boundary becomes complicated, but we overlook these situations in Fig. 3.2(b), since such ranges of the external force are extremely limited.

40

vary slowly but periodically. In the same way that a singular point represents a periodic solution of the initial system, a limit cycle represents a stationary oscillation which is affected by amplitude and phase modulation.

The closed trajectory C is said to be orbitally stable if, given $\epsilon > 0$, there is $\eta > 0$ such that, if R is a representative point of another trajectory which is within a distance η from C at time t_0, then R remains within a distance ϵ from C for $t > t_0$. If no such η exists, C is orbitally unstable. Moreover if C is orbitally stable and, in addition, if the distance between R and C tends to zero as t increases, C is said to be asymptotically orbitally stable [19, 32].

The stability of a limit cycle can be tested by making use of Poincaré's criterion for orbital stability. This stability criterion is the following inequality [19];

$$\oint_C \left(\frac{\partial X_n}{\partial x_n} + \frac{\partial Y_n}{\partial y_n} \right) dt < 0. \tag{3.23}$$

We proceed to establish the existence of a limit cycle for the derived autonomous systems when the external force is given outside the regions of frequency entrainment. In such a case it follows from a careful consideration that there is only one singular point in the $x_n y_n$ plane. Furthermore, this singular point is identified as an unstable focus. This means that any representative point starting near this singularity moves away from it with increasing t; in fact there is an ellipse containing the focus in its interior with the property that all representative points cross it moving from its interior to its exterior as t increases. On the other hand, all integral curves of the autonomous systems remain, as t increases, within a circle of sufficiently large radius. This follows at once from the form of the autonomous systems (3.8), (3.10), (3.11), (3.12) and (3.13), since we have approximately for large x_n and y_n: $dx_n/dt = -(\mu/2)r_n^2 x_n$ and $dy_n/dt = -(\mu/2)r_n^2 y_n$, so that $dy_n/dx_n = y_n/x_n$. This means that the integral curves are approximately rays through the origin and that a representative point on one of them moves toward the origin as t increases. Thus there is a ring-shaped domain bounded on the outside by this circle and on the inside by a small ellipse which is free from singular points and has the property that any solution curve which starts inside it remains there as t increases. The theorem of Poincaré and Bendixson can therefore be applied to establish the existence of at least one limit cycle [10, 32-33].

Thus far, the existence of a limit cycle for the derived autonomous systems is established. However, it is in general not easy to obtain any further information (number, location, size and shape) about such a limit cycle. In order to determine the limit cycle precisely, we are compelled to resort to numerical or graphical means.

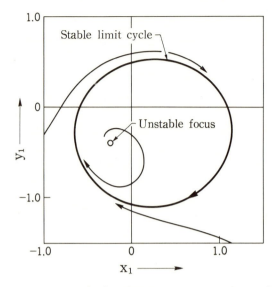

Fig. 3.3 Limit cycle for the autonomous system (3.8),
the parameters being $\mu = 0.15$, $\beta = 4/3$, $\gamma = 4/3$,
$B = 0.2$ and $\nu = 1.1$.

Let us give an example of a limit cycle when the amplitude B and the frequency ν of the external force are prescribed close to the regions of entrainment. The system parameters considered are the same as those in Fig. 3.2.

By putting the external force just outside the region of harmonic entrainment, i.e.,

$$B = 0.2 \quad \text{and} \quad \nu = 1.1,$$

we obtain an example of a limit cycle of Eqs. (3.8). Figure 3.3 shows the limit cycle of this case, which is obtained by carrying out numerical integration both from the inside and from the outside of the limit cycle. The result establishes that the limit cycle is asymptotically orbitally stable. The periodic solution of Eqs. (3.8) represented by the limit cycle of Fig. 3.3 is given by

$$
\begin{aligned}
x_1(t) = \varphi_1(t) = \ & 0.09 + 0.19\sin\theta \quad -\, 0.81\cos\theta \\
& - 0.19\sin 2\theta + 0.06\cos 2\theta \\
& + 0.04\sin 3\theta + 0.04\cos 3\theta + \cdots \\
y_1(t) = \psi_1(t) = & -0.42 + 0.74\sin\theta \quad +\, 0.17\cos\theta \\
& - 0.06\sin 2\theta - 0.17\cos 2\theta \\
& - 0.03\sin 3\theta + 0.04\cos 3\theta + \cdots
\end{aligned}
\tag{3.24}
$$

where

$$\theta = 0.0841 \cdots \times (t - h) \quad \text{and} \quad h : \text{ an arbitrary constant.}$$

The period τ of this solution is approximately $74.68 (= 2\pi/\nu \times 13.07 \cdots)$. The variational equation for the periodic state (3.24) is a linear differential equation with periodic coefficients,

$$\frac{d\xi}{dt} = p_{11}(t)\xi + p_{12}(t)\eta, \qquad \frac{d\eta}{dt} = p_{21}(t)\xi + p_{22}(t)\eta \qquad (3.25)$$

where

$$p_{11}(t) = \frac{\mu}{2}\{1 - 3\varphi_1^2(t) - \psi_1^2(t)\}, \quad p_{12}(t) = \frac{\mu}{2}\{-\sigma_1 - 2\varphi_1(t)\psi_1(t)\}$$

$$p_{21}(t) = \frac{\mu}{2}\{\sigma_1 - 2\varphi_1(t)\psi_1(t)\}, \qquad p_{22}(t) = \frac{\mu}{2}\{1 - \varphi_1^2(t) - 3\psi_1^2(t)\}.$$

As the Eq. (3.8) is autonomous and its periodic solution $(\varphi_1(t), \psi_1(t))$ is not a state of equilibrium, the linear system (3.25) necessarily has one characteristic multiplier equal to unity, or equivalently one characteristic exponent equal to zero. The other characteristic multiplier ρ is given by [9, 19]

$$\rho = \exp\left[\mu \int_0^\tau \{1 - 2\varphi_1^2(t) - 2\psi_1^2(t)\}dt\right]$$

$$= \exp\left[\mu\tau(1 - 2\bar{r}_1^2)\right] = \exp(-8.20) = 0.000275 \qquad (3.26)$$

where

$$\bar{r}_1^2 = \frac{1}{\tau}\int_0^\tau \{\varphi_1^2(t) + \psi_1^2(t)\}dt = 0.866.$$

This confirms that the limit cycle is strongly stable.[§] The magnitude $\bar{r}_1$ is sometimes called the normalized r.m.s (root mean square) amplitude of an almost periodic oscillation.

(f) Transition between entrained oscillations and almost periodic oscillations

Let us here briefly explain the transition between entrained oscillations and almost periodic oscillations when the external force is varied across the boundary of harmonic entrainment. As stated above, the boundary of harmonic entrainment is given by the conditions (3.21). The second condition $q > 0$ applies in the case where the detuning σ_1 and consequently the amplitude B are comparatively small. On the other hand, the first condition $p > 0$ applies in the case where σ_1 and B are large.

First, let us suppose the case in which almost periodic oscillations are sustained for comparatively small detuning σ_1, and from this state the frequency

---

[§]The strong stability of limit cycles is defined by the condition $\rho < 1$; or one of the characteristic exponents of the Eqs. (3.25) is less than zero [34].

ν and/or the amplitude B are varied across the boundary $q = 0$ into the region of harmonic entrainment. When the external force reaches the boundary $q = 0$, a higher order singularity appears suddenly on the limit cycle representing the almost periodic oscillation. This singularity is a coalesced node-saddle point, and as the external force enters into the region of harmonic entrainment, the singular points separate from one another; i.e., stable node and saddle. Thus the synchronized oscillation is established.

Second, let us consider the case in which almost periodic oscillations are sustained for large detuning σ_1, and from this state the external force is brought inside the region of harmonic entrainment through the boundary $p = 0$. In this situation limit cycles representing almost periodic oscillations become small in size, and when the external force is prescribed on the boundary $p = 0$ the limit cycle shrinks to a point; at the same time the unstable focus inside the limit cycle turns into a stable focus. Thus the synchronized periodic oscillation appears.

The difference in the above descriptions can be explained as follows. An almost periodic oscillation may be considered as a combination of two components, i.e., the free oscillation with the natural frequency of the system and the forced oscillation with the driving frequency. When the detuning σ_1 is small, the forced oscillation is not predominant, and the free oscillation is entrained by the driving frequency. On the other hand, for large detuning σ_1, the forced oscillation is predominant and the free oscillation is suppressed by the forced one. It should be added that no hysteresis phenomena are observed while varying the external force across the boundary of entrainment for the above two situations, both for small and for large detuning. However, for intermediate values of the detuning σ_1; i.e., $1/2 \leq |\sigma_1| \leq 1/\sqrt{3}$ (see Fig. 3.1), complicated phenomena occur. As the ranges of the frequency ν and the amplitude B are extremely narrow for this case and consequently the original non-autonomous equation (3.2) cannot be treated by analog computer, we do not discuss this problem here. Quite similar phenomena are observed for rather wide ranges of the external force in forced self-oscillatory systems with nonlinear restoring term, and the phenomena will be discussed in the following chapters 4 and 5.

3.4 Analog and Digital Computer Analyses

The results obtained in the preceding section are compared with the solutions of the original Eq. (3.1) or (3.2) obtained by using analog and digital computer.

The block diagram of Fig. 3.4 shows an analog computer setup for the solution of Eq. (3.2), in which the system parameters μ, β and γ are set equal

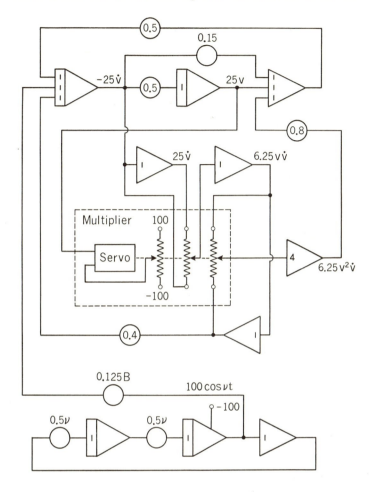

Fig. 3.4 Block diagram of an analog computer setup for the
solution of Eq. (3.2), the system parameters being
$\mu = 0.15$, $\beta = 4/3$ and $\gamma = 4/3$.

to the values as given above; i.e., $\mu = 0.15$, $\beta = 4/3$ and $\gamma = 4/3$. It is confirmed that the regions of frequency entrainment thus obtained agree well with that of Fig. 3.2.

Here, let us investigate the behavior of the images under the mapping T of the $v\dot{v}$ plane into itself, which is defined by Eqs. (3.5), or (3.7), in much the same way as described in Sec. 1.5.

As we have seen in Sec. 3.3(e), when the amplitude B and the frequency ν of the external force are given outside the regions of entrainment, the corresponding autonomous system possesses at least one stable limit cycle. However, it

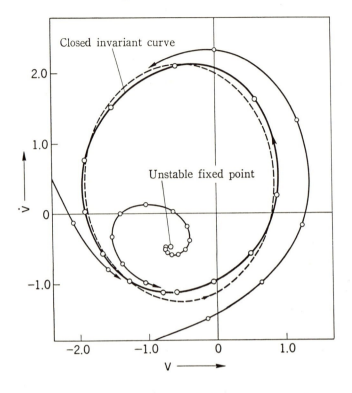

Fig. 3.5 Invariant closed curve of the mapping for Eqs. (3.5),
the parameters being $\mu = 0.15$, $\beta = 4/3$, $\gamma = 4/3$,
$B = 0.2$ and $\nu = 1.1$.

is tacitly assumed that to the limit cyle there actually corresponds an almost
periodic solution of the initial system (3.7). It is known that, when the limit
cycle is strongly stable, there exists a unique correspondence between the limit
cycle of the autonomous system (3.8) and the simple closed curve which is invariant under the mapping T defined by Eqs. (3.7) and differs little from the
limit cycle provided μ is sufficiently small [34]. The closed curve invariant under
the mapping T is called an invariant closed curve. The movement of images on
the invariant closed curve is characterized by the rotation number, which mea-
sures the average advance of an image under the mapping T. It is ascertained
that, when the rotation number is irrational and the invariant closed curve is
sufficiently smooth, there exists almost periodic solutions in Eqs. (3.7), or in
Eqs. (3.5) [32].

Figure 3.5 shows an example of an invariant closed curve of the mapping
T defined by Eqs. (3.5). The system parameters and the external force are
taken equal to the values as used in Fig. 3.3; i.e., $\mu = 0.15$, $\beta = 4/3$, $\gamma = 4/3$,

$B = 0.2$ and $\nu = 1.1$. The successive images $P_n(v(2n\pi/\nu), \dot{v}(2n\pi/\nu))$ ($n = 1$, 2, 3, $\cdots$) of the initial point $P_0(v(0), \dot{v}(0))$ under the mapping T are obtained both from the inside and from the outside of the invariant closed curve. The small circles on the curves represent examples of images under the mapping T, and they are transformed successively to the points that follow in the direction of the arrows. (It should be noted that the solid curves connecting the dots approaching the invariant closed curve do not represent solutions of Eq. (3.2), but are merely intended to guide the eye.) The period required for the image to complete one revolution along the invariant closed curve of Fig. 3.5 is $11.37\cdots$ times the period of the external force. This implies that the rotation number associated with the invariant closed curve is $0.912\cdots$.[¶] The closed curve drawn as a dashed line indicates the limit cycle of Fig. 3.3 converted through the relation (3.6).

[¶]The approximate solution (3.3) or the first equation of (3.6) may be represented as

$$v(t) = a_0 r_1(t) \sin[\nu t + \theta(t)]$$

where $\theta(t) = \tan^{-1} y_1(t)/x_1(t)$.

As the limit cycle of Fig. 3.3 contains the origin $x_1 = y_1 = 0$ in its interior and the representative point moves in the clockwise direction, the phase angle of the oscillation lags 2π radians when the representative point makes one revolution along the limit cycle. Therefore the rotation number associated with the invariant closed curve of Fig. 3.5 is approximated by $1 - 1/11.37 \simeq 0.912$.

4. HYSTERETIC TRANSITION FOR INTERMEDIATE VALUES OF THE DETUNING

4.1 Van der Pol/Duffing Mixed Type Equation

In the preceding chapter we treated the problem in which the damping of the system was nonlinear. In this chapter we consider a problem in which both the damping and the restoring force are nonlinear; that is, a system governed by the van der Pol/Duffing mixed type equation

$$\frac{d^2v}{dt^2} - \mu(1 - \gamma v^2)\frac{dv}{dt} + v^3 = B\cos\nu t \tag{4.1}$$

where μ is a small positive constant and γ is positive also, and $B\cos\nu t$ represents a forcing term.

In the preceding chapter we overlooked the case where both entrained oscillations and almost periodic oscillations occur even for the same value of the external force, since such a range of the external force is extremely narrow in Eq. (3.1). On the other hand in the system described by Eq. (4.1) the response curves are skewed by the nonlinear restoring force as will be seen later. From this it may be conjectured that such a range of the external forces becomes broader than before. Therefore we consider a particular case in which two types of steady-state responses occur depending only on different value of the initial conditions. Special attention is directed toward the transition between entrained oscillations and beat oscillations which occurs under such circumstances.* The method of analysis is much the same as we have used in the foregoing chapter.

4.2 Response Curves and the Region of Harmonic Entrainment

When no external force is applied to the system, a free oscillation is self-excited. To find the ultimate amplitude and frequency of the oscillation, we put, to a first approximation,

$$v(t) = a_0\cos\omega_0 t \tag{4.2}$$

---

*In the original version of the text, the term almost periodic oscillation is used for beat oscillation, which occurs when the external force is given outside the regions of entrainment. In the present version, however, we use the term beat oscillations instead of almost periodic oscillations.

and substitute this into Eq. (4.1) with $B = 0$. Then, equating the coefficients of the terms containing $\sin \omega_0 t$ and $\cos \omega_0 t$ separately to zero, we obtain

$$a_0^2 = \frac{4}{\gamma}, \qquad \omega_0^2 = \frac{3}{4}a_0^2 = \frac{3}{\gamma}. \tag{4.3}$$

We see that the natural frequency ω_0 of the system is proportional to the amplitude a_0 of the free oscillation. This results from the presence of the nonlinear restoring term.

When the external force with frequency ν nearly equal to ω_0 is present, either an entrained harmonic oscillation or a beat oscillation which develops from harmonic entrainment may occur depending on the value of B. Hence we assume an approximate solution of the form

$$v(t) = b_1(t) \sin \nu t + b_2(t) \cos \nu t \tag{4.4}$$

where $b_1(t)$, $b_2(t)$ become constants for entrained oscillations and slowly varying function of the time t for beat oscillations.

Here let us derive an autonomous equation which determines the amplitudes, $b_1(t)$ and $b_2(t)$ in Eq. (4.4). Substituting Eq. (4.4) into (4.1) and equating the coefficients of the terms containing $\cos \nu t$ and $\sin \nu t$ separately to zero leads to

$$
\begin{aligned}
\frac{dx_1}{dt} &= \frac{\mu}{2}\left[(1 - r_1^2)x_1 - \sigma_1 y_1 + \frac{B}{\mu \nu a_0}\right] \equiv X_1(x_1, y_1) \\
\frac{dy_1}{dt} &= \frac{\mu}{2}\left[\sigma_1 x_1 + (1 - r_1^2)y_1\right] \qquad\quad \equiv Y_1(x_1, y_1)
\end{aligned}
\tag{4.5}
$$

where

$$x_1 = \frac{b_1}{a_0}, \qquad y_1 = \frac{b_2}{a_0}, \qquad r_1^2 = x_1^2 + y_1^2$$

$$a_0 = \sqrt{\frac{4}{\gamma}}, \quad \omega_0 = \sqrt{\frac{3}{\gamma}}, \quad \sigma_1 = \frac{\omega_0^2 r_1^2 - \nu^2}{\mu \nu} \text{ (detuning)}.$$

In deriving the autonomous equations (4.5), the following assumptions are used:

1. The amplitudes $b_1(t)$ and $b_2(t)$ are slowly varying functions of t; therefore, $d^2 b_1/dt^2$ and $d^2 b_2/dt^2$ are neglected.

2. Since μ is a small quantity, $\mu db_1/dt$ and $\mu db_2/dt$ are also discarded.

The equations have the same form as Eqs. (3.8), but it should be noted that the definition of the detuning σ_1 differs from that in Sec. 3.3(a). In the present case, the detuning is a function of both the amplitude r_1 and the driving frequency ν.

This is a reasonable result since the natural frequency of the system is modified to $\omega_0 r_1$. Equations (4.5) serve as the fundamental equations in studying beat oscillations as well as entrained oscillations. The entrained harmonic oscillation is obtained by putting $dx_1/dt = 0$ and $dy_1/dt = 0$, i.e.,

$$X_1(x_1, y_1) = 0, \qquad Y_1(x_1, y_1) = 0. \tag{4.6}$$

The above equations can be combined to give a single equation for r_1,

$$[(1 - r_1^2)^2 + \sigma_1^2]r_1^2 = \left(\frac{B}{\mu\nu a_0}\right)^2. \tag{4.7}$$

Once r_1 is determined, x_1 and y_1, that is, the coordinates of the singular point, are found to be

$$x_1 = -\frac{\mu\nu a_0}{B}(1 - r_1^2)r_1^2, \qquad y_1 = \frac{\mu\nu a_0}{B}\sigma_1 r_1^2. \tag{4.8}$$

Figure 4.1 shows the response curves as given by Eq. (4.7) for the following values of the system parameters;

$$\mu = 0.2 \qquad \text{and} \qquad \gamma = 8.$$

One sees that the response curves in the case of nonlinear restoring force could be thought of as arising from those for the linear case (see Fig. 3.1) by bending the latter to the right. As a consequence of this bending, it can be seen in Fig. 4.1 that for many values of B and ν three different response amplitudes are possible.

Proceeding analogously as in Sec. 3.3(c), let us investigate the stability of the equilibrium state by considering the behavior of small variations ξ and η from the singular point and determine whether these deviations approach zero or not with increase of t. The variational equation is

$$\frac{d\xi}{dt} = a_1\xi + a_2\eta, \qquad \frac{d\eta}{dt} = b_1\xi + b_2\eta \tag{4.9}$$

with

$$a_1 = (\partial X_1/\partial x_1)_0, \ a_2 = (\partial X_1/\partial y_1)_0, \ b_1 = (\partial Y_1/\partial x_1)_0, \ b_2 = (\partial Y_1/\partial y_1)_0$$

where $(\partial X_1/\partial x_1)_0, \ \cdots, \ (\partial Y_1/\partial y_1)_0$ stand for $\partial X_1/\partial x_1, \ \cdots, \ \partial Y_1/\partial y_1$ at the singular point. By making use of the Routh-Hurwitz criterion, the conditions

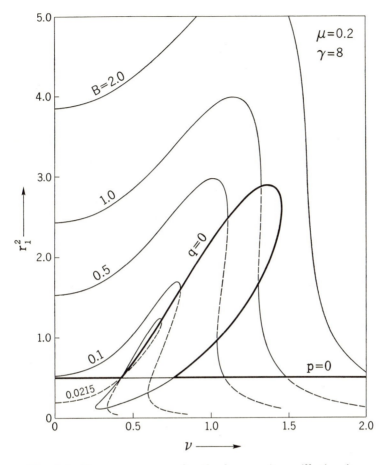

Fig. 4.1 Response curves for the harmonic oscillation in
the system described by Eq. (4.1), the system
parameters being $\mu = 0.2$ and $\gamma = 8$.

for stability of the singular point are given by

$$p = -(a_1 + b_2) = \mu(2r_1^2 - 1) > 0$$
$$q = a_1 b_2 - a_2 b_1 = \frac{\mu^2}{4}\left[(1 - r_1^2)(1 - 3r_1^2) + \sigma_1^2 + 2\frac{\omega_0^2}{\mu\nu}\sigma_1 r_1^2\right] > 0. \tag{4.10}$$

The stability limits $p = 0$ and $q = 0$ are also shown in Fig. 4.1. Hence the dashed portions of the response curves are unstable. It is readily verified that the vertical tangencies of the response curves lie on the stability limit $q = 0$.

From the above results, the region of harmonic entrainment is obtained in the $B\nu$ plane as illustrated in Fig. 4.2. In this plane each point (ν, B) cor-

51

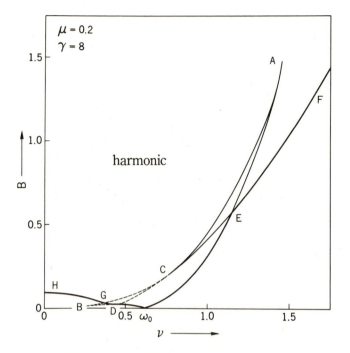

Fig. 4.2 Region of harmonic entrainment in the system
described by Eq. (4.1), the system parameters being
$\mu = 0.2$ and $\gamma = 8$.

responds to a particular pattern of integral curves; i.e., a phase portrait of
Eqs. (4.5) in the $x_1 y_1$ plane. The number and character of the singularities are
prescribed by the location of the point (ν, B). The boundary curve $BD\omega_0 EA$
corresponds to the upper portion of $q = 0$ in Fig. 4.1, and $BGCA$ to the
lower portion of $q = 0$. For points on these boundaries, phase portraits have a
higher-order singularity corresponding to a point on the stability limit $q = 0$.
The boundary curve $HDCF$ corresponds to the boundary $p = 0$. It touches the
curve BGA at the point C and $B\omega_0$ at D. The points C and D correspond to
the intersections of $q = 0$ with $p = 0$ in Fig. 4.1. For points on the curve CD,
the singularity on $p = 0$ is a saddle point within the unstable region $q < 0$, and
hence, this curve has no practical significance. On the other hand, for points
on the curves HD and CF, the singularity on $p = 0$ is a higher-order singularity
representing the transition between a stable and unstable focus.

From the above considerations, we see that when B and ν are in the region
represented by the curvilinear triangle $\omega_0 AB$, Eqs. (4.5) have three singularities,
one of which is a saddle point. Outside this region, there is only one stable
or unstable singularity, according to whether the point in question is above or

52

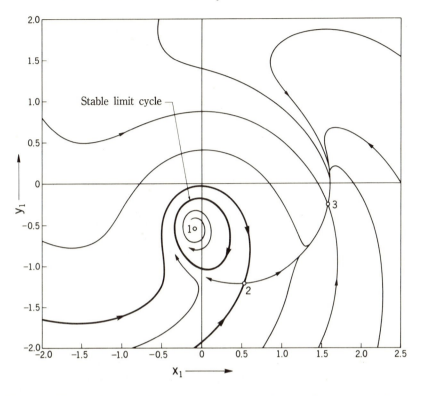

Fig. 4.3 Example of a phase portrait in which singularities and limit
cycle coexist, the parameters being $\mu = 0.2$, $\gamma = 8$,
$B = 0.35$ and $\nu = 1.0$.

below the boundaries HG and EF. In the regions represented by the curvilinear triangles ACE and BDG there are three singularities, one of which is a saddle point and the remaining two are stable singularities in the region ACE, and unstable singularities in BDG. The region above the line $HD\omega_0 EF$ (drawn as a thick line in the figure), therefore, becomes the region of harmonic entrainment since there is at least one stable singularity in the phase portrait.

4.3 Hysteretic Transition between Entrained Oscillations and Beat Oscillations

From the results above, we see that if the amplitude B and frequency ν of the external force are prescribed to the right of the region of harmonic entrainment $(\nu > \omega_0)$, Eqs. (4.5) have only an unstable singularity. It will also be seen that, for large values of x_1 and y_1, the representative point on the integral curve will cross the family of concentric circles centered at the origin from the outside to

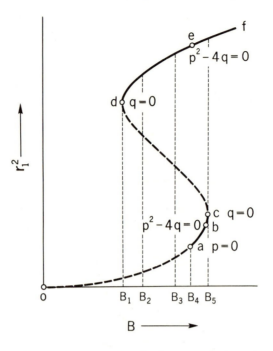

Fig. 4.4 Example of amplitude characteristics when the
driving frequency ν is kept constant at $\nu = 0.9$.

the inside as t increases. Hence the existence of a stable limit cycle in the $x_1 y_1$ plane may be concluded, which corresponds to a beat oscillation. However, even if B and ν are given inside the region of harmonic entrainment, Eqs. (4.5) can possess a stable limit cycle in addition to a stable singular point for certain values of B and ν.

Figure 4.3 shows an example of a phase portrait of such a case. The amplitude B and the frequency ν of the external force are given by

Table 4.1 Classification of singular points along
the amplitude characteristic of Fig. 4.4

Interval	p	q	$p^2 - 4q$	Classification
oa	−	+	−	Unstable focus
ab	+	+	−	Stable focus
bc	+	+	+	Stable node
cd	+	−	+	Saddle
de	+	+	+	Stable node
ef	+	+	−	Stable focus

54

Some Problems in the Theory of Nonlinear Oscillations

$$B = 0.35 \qquad \text{and} \qquad \nu = 1$$

which are located in the region of harmonic entrainment (see Fig. 4.2). The other system parameters are the same as in Fig. 4.1. We see in the figure a stable limit cycle which encircles the unstable focus 1. The integral curves (drawn as a thick line in the figure) that approach the saddle point 2 form a separatrix which divides the whole plane into two regions; in one all integral curves tend to the limit cycle, while in the other they tend to the nodal point 3. Thus both beat oscillation and entrained oscillation may occur in this system, which one it will be depending on the initial condition. Let us here consider the transitions between entrained oscillations and beat oscillations which occur under such circumstances.

Figure 4.4 shows an example of the amplitude characteristic (Br_1^2-relation) when ν is kept constant ($= 0.9$). By making use of the stability conditions (4.10) we classified the singular points along the amplitude characteristic illustrated in Fig. 4.4. The results are listed in Table 4.1.

The occurrence and extinction of a limit cycle, as B varies along this characteristic, will most easily be understood by plotting the phase portraits of Eqs. (4.5) for various values of B. The results are shown in Fig. 4.5. When B is given below B_1 of Fig. 4.4, a limit cycle exists which encircles an unstable focal point. This state is shown in Fig. 4.5(a). A coalesced singularity appears at $B = B_1$. Further increase in B results in the separation of this higher-order singular point into two simple singular points, resulting in the coexistence of a stable limit cycle and a stable node. When $B = B_2$, there exists a closed integral curve starting from the saddle point and coming back to the same point, as shown in Fig. 4.5(d). The limit cycle disappears when B is increased beyond B_2. However, when B reaches B_3 the integral curves once again show the same behavior as in Fig. 4.5(d). A limit cycle appears once more for $B_3 < B < B_4$ and shrinks up to a stable focus at $B = B_4$. For values of B between B_4 and B_5, there exist two distinct stable singularities, corresponding to resonant and non-resonant oscillations, but there is no limit cycle. The coalescence of singularities occurs at $B = B_5$ and then this higher-order singularity disappears. Hence it is concluded that a beat oscillation occurs for $B < B_2$ and $B_3 < B < B_4$, a resonant oscillation for $B_1 < B$, and a non-resonant oscillation for $B_4 < B < B_5$.

Here we introduce the mean-square amplitude of the beat oscillation defined by

$$\bar{r}_1^2 = \frac{1}{\tau} \oint_C r_1^2 dt \qquad (4.11)$$

where C is an integration path taken around the limit cycle and τ is the period

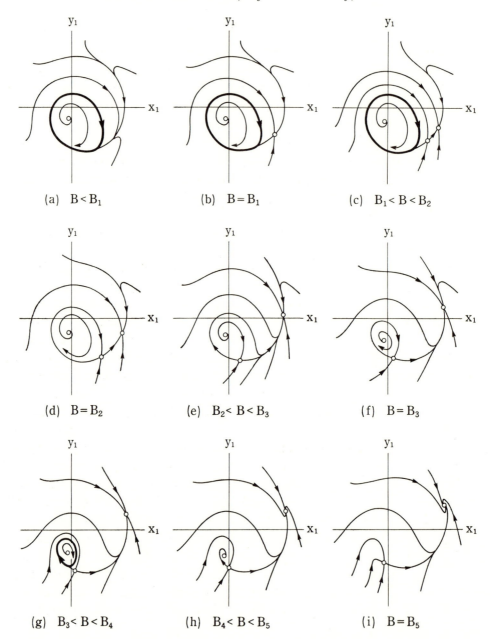

Fig. 4.5 Various phase portraits of Eqs. (4.5) showing the transition between
entrained oscillations and beat oscillations.

for the representative point $(x_1(t), y_1(t))$ to complete one revolution along the
limit cycle. This mean square amplitude $\bar{r}_1^2$ of the beat oscillation is plotted by

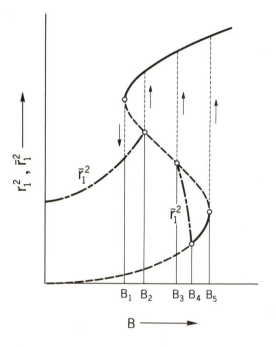

Fig. 4.6 Amplitude characteristics with mean-square
amplitude of the beat oscillations.

uneven dashed lines in Fig. 4.6. The arrows in the figure show the transitions
between these oscillations. Thus we see that between B_1 and B_2 there exists
hysteresis and at B_3 and B_5 jump phenomena take place from the lower to the
higher branch.

It should also be mentioned that, for other values of the external force parameters, the situation may somewhat be different from that mentioned above.
Under certain conditions (for example, for different values of ν), a beat oscillation occurs for any value of B below B_4, that is, the two separate uneven
dashed lines become one; or a beat oscillation may occur only for values below
B_1 or between B_3 and B_4. In the latter case no hysteresis between B_1 and B_2
results, since the coalescence of singularities appears on the limit cycle.

In the above we investigated the right hand transitional region as given by
$\nu > \omega_0$, but for the left hand region $\nu < \omega_0$, particularly in the neighborhood
of the curvilinear triangle BDG in Fig. 4.2, some complicated phenomena may
occur [35-36]. However, we shall not enter into this problem here, because such
a region of the external force parameters is extremely narrow. An interesting
example of a transition for such a situation will be given in the last chapter of
this text.

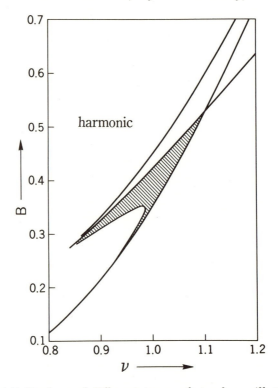

Fig. 4.7 Regions of different types of steady oscillations
obtained by analog computer analysis.

4.4 Analog Computer Analysis

In the preceding section two types of steady-state responses and the transition between them were investigated on the basis of the derived autonomous system (4.5). Here let us examine these results by using an analog computer for the original non-autonomous system.

The solutions of the original differential equation (4.1), i.e.,

$$\frac{d^2v}{dt^2} - \mu(1 - \gamma v^2)\frac{dv}{dt} + v^3 = B\cos\nu t$$

are sought for rather restricted ranges of B and ν, concentrating our attention on the hysteretic transition for the case in which the driving frequency ν is greater than the natural frequency ω_0 of the system. The system parameters are the same as in Secs. 4.2 and 4.3, i.e., $\mu = 0.2$ and $\gamma = 8$. The regions of different steady states thus obtained are shown in Fig. 4.7. The lines in the figure show the boundaries of entrainment, or boundaries of different types of oscillations. The hatched area indicates the presence of a beat oscillation in addition to an entrained oscillation. The other domains of the different types of oscillations are the same as already explained in Fig. 4.2 and should be self-explanatory. The result concerning hysteresis is also confirmed.

5. TRANSITIONS OF SINGULARITIES AND LIMIT CYCLES

5.1 Autonomous System Derived from Rayleigh/Duffing Mixed Type Equation

In the preceding chapter, we investigated the phenomenon of hysteretic transition which occurs in the self-oscillatory system with external force described by a second-order non-autonomous equation, the van der Pol/Duffing mixed type equation. The procedure was: first, the non-autonomous equation was approximated by an autonomous equation by applying the averaging method. Then, the singular points and the limit cycles of the derived autonomous system were investigated. Finally, by using an analog computer the results for the autonomous system were compared with the solutions of the original non-autonomous system.

In the present chapter, however, ignoring the applicability of the results for an autonomous system to a non-autonomous one, we consider the transitions of singularities and limit cycles of an autonomous system which determines the amplitudes of approximate steady states of a self-oscillatory system with external force. The reason is as follows: the parameter ranges treated in the following analysis are rather restricted and narrow; moreover the phenomenon itself in the circumstances is so complicated that it cannot be expected that analog-computer experiments can bring reliable solutions of the original non-autonomous system and that the phenomenon can be approximated by a simple form of solutions as was used in the preceding chapter.

Let us consider the autonomous system

$$
\begin{aligned}
\frac{dx}{dt} &= \frac{\mu}{2}\left[(1 - \gamma\nu^2 r^2)x - \sigma y + \frac{B}{\mu\nu a_0}\right] \equiv X(x,\,y) \\
\frac{dy}{dt} &= \frac{\mu}{2}\left[\sigma x + (1 - \gamma\nu^2 r^2)y\right] \qquad\quad \equiv Y(x,\,y)
\end{aligned}
\tag{5.1}
$$

where

$$
r^2 = x^2 + y^2, \quad a_0^2 = \frac{4}{3\gamma}, \quad \omega_0^2 = \frac{1}{\gamma}, \quad \sigma = \frac{\omega_0^2 r^2 - \nu^2}{\mu\nu} \ \text{(detuning)}.
$$

The equations (5.1) arise as a derived autonomous system for the components in a phase plane rotating with frequency ν of the forced oscillations in a self-oscillatory system with external force governed by the Rayleigh/Duffing mixed

60

type equation

$$\frac{d^2v}{dt^2} - \mu\left[1 - \gamma^2\left(\frac{dv}{dt}\right)^2\right]\frac{dv}{dt} + v^3 = B\cos\nu t, \quad 0 < \mu \ll 1. \tag{5.2}$$

By assuming an approximate solution for the above equation (5.2) of the form

$$v(t) = b_1(t)\sin\nu t + b_2(t)\cos\nu t \tag{5.3}$$

and by applying the harmonic balance method under the same assumptions made on $b_1(t)$ and $b_2(t)$ as in the preceding chapter, the equations (5.1) are derived with the relations $x = b_1(t)/a_0$ and $y = b_2(t)/a_0$. The constants a_0 and ω_0 represent the values of the first approximation to the amplitude and frequency of the free oscillation, respectively. In the systems (5.1) and (5.2), μ and γ are system parameters; B and ν represent the amplitude and frequency of the external force; that is, the same notations as in the preceding chapter are used.

The system (5.2) also describes the phenomenon of frequency entrainment. Before giving an example of transition of singularities and limit cycles, let us show an outline of the region of harmonic entrainment.

5.2 Response Curves and the Region of Harmonic Entrainment

The region of harmonic entrainment is the region in the $B\nu$ plane such that, if the amplitude B and the frequency ν of the external force are given in the region, the autonomous system (5.1) possesses at least one stable singularity.

The coordinates of the singular point are given by

$$x = -\frac{\mu\nu a_0}{B}(1 - \gamma\nu^2 r^2)r^2, \qquad y = \frac{\mu\nu a_0}{B}\sigma r^2 \tag{5.4}$$

where r^2 is determined by the relation

$$[(1 - \gamma\nu^2 r^2)^2 + \sigma^2]r^2 = \left(\frac{B}{\mu\nu a_0}\right)^2. \tag{5.5}$$

The conditions for stability of the singular point (5.4) may also be derived by the same procedure as used in the preceding chapters. The results are,

$$p = \mu(2\gamma\nu^2 r^2 - 1) > 0$$

$$q = \frac{\mu^2}{4}\left[(1 - \gamma\nu^2 r^2)(1 - 3\gamma\nu^2 r^2) + \sigma^2 + 2\frac{\omega_0^2}{\mu\nu}\sigma r^2\right] > 0. \tag{5.6}$$

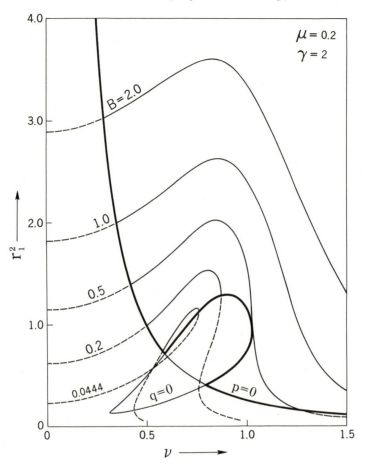

Fig. 5.1 Response curves for the system described by Eqs. (5.1),
the system parameters being $\mu = 0.2$ and $\gamma = 2$.

Figure 5.1 shows the response curves given by the relation (5.5) for the following
values of the system parameters;

$$\mu = 0.2 \qquad \text{and} \qquad \gamma = 2.$$

The stability limits $p = 0$ and $q = 0$ given by Eqs. (5.6) are also shown in the
figure. The dashed portions of the resonance curves are unstable. One sees that
Fig. 5.1 is similar to Fig. 4.1 except for the shape of the stability limit $p = 0$.
From this result the region of harmonic entrainment is reproduced in the $B\nu$
plane as illustrated in Fig. 5.2, which is also analogous to Fig. 4.2. In the figure
the curvilinear triangle BDG corresponding to that of Fig. 4.2 is broader than
before. Therefore, we can more readily deal with the behavior of integral curves

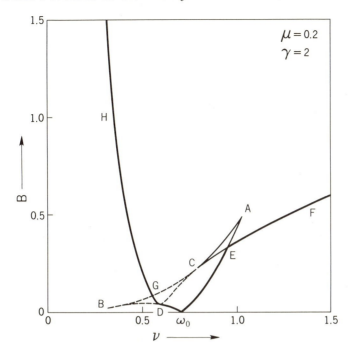

Fig. 5.2 Region of harmonic entrainment in the system
described by Eqs. (5.1), the system parameters being
$\mu = 0.2$ and $\gamma = 2$.

when the external force is given in the vicinity of this region, or to the left of
the frequency ω_0 of the free oscillation.

5.3 Transition of Singularities and Limit Cycles

Let us first show an example of a phase portrait having three singularities
and two limit cycles. The special case we have in mind is that specified by the
following values of the external force parameters

$$B = 0.065 \qquad \text{and} \qquad \nu = 0.57$$

which is located in the curvilinear triangle BDG of Fig. 5.2. The other system
parameters are the same as in Figs. 5.1 and 5.2. The phase portrait of Eqs. (5.1)
for the above parameters is plotted in Fig. 5.3. As illustrated in the figure, there
are three singularities in this case. We see in the figure an unstable limit cycle
which encircles the stable focus 3, and outside this a stable limit cycle enclosing
three singular points. The unstable limit cycle divides the whole plane into two
regions; inside this limit cycle all integral curves tend to the stable focus 3,

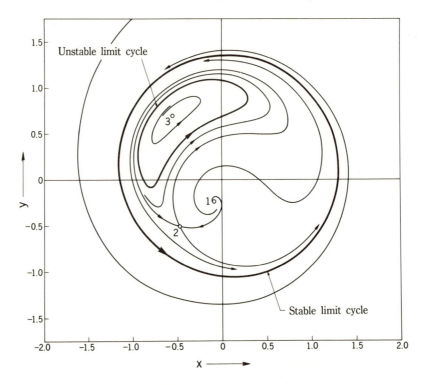

Fig. 5.3 Example of a phase portrait in which singularities and limit
cycles coexist, the parameters being $\mu = 0.2$, $\gamma = 2$,
$B = 0.065$ and $\nu = 0.57$.

while outside they tend to the stable limit cycle. Thus both the stable singular
point and the limit cycle coexist in this system.

Here, let us proceed to the main subject of the transitions of singularities
and limit cycles of Fig. 5.3 when B is varied. By substituting $\nu = 0.57$ into Eq.
(5.5), the Br^2 relation is calculated and plotted in Fig. 5.4. As in the case of
Sec. 4.3, the singularities are classified and the results are listed in Table 5.1.

Figure 5.5 shows the phase portraits of Eqs. (5.1) for various values of B.
The phase portrait of Fig. 5.3 appears again as Fig. 5.5(d). When B is decreased
the unstable limit cycle diminishes in size and shrinks up to the unstable focus
when B_2 of Fig. 5.4 is reached. This state is shown in Fig. 5.5(c). Further
decrease in B brings about the coalescence of singular points at $B = B_1$ and
then this higher-order singularity disappears (see Figs. 5.5(a) and (b)). When
B is increased from Fig. 5.5(d) the unstable limit cycle grows large, and at
$B = B_3$ a higher-order singularity is formed by the coalescence of two singular
points at a location lying between two limit cycles. Two limit cycles persist

64

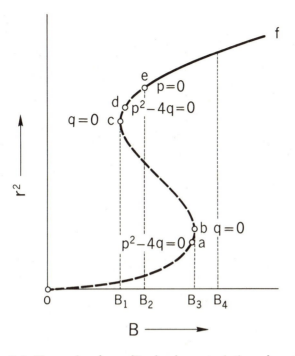

Fig. 5.4 Example of amplitude characteristics when the
driving frequency is kept constant at $\nu = 0.57$.

for values of B between B_3 and B_4 coexisting with one stable singular point;
when B reaches B_4 the two limit cycles coincide and form a higher-order limit
cycle which is semi-stable (Fig. 5.5(h)), which then disappears leaving only one
stable focus.

Hence it is concluded that a stable limit cycle appears for $B < B_4$, an
unstable limit cycle for $B_2 < B < B_4$ and a stable singular point for $B_2 < B$
in this particular case.

Table 5.1 Classification of singular points along
the amplitude characteristic of Fig. 5.4

Interval	p	q	$p^2 - 4q$	Classification
oa	−	+	−	Unstable focus
ab	−	+	+	Unstable node
bc	−	−	+	Saddle
cd	−	+	+	Unstable node
de	−	+	−	Unstable focus
ef	+	+	−	Stable focus

65

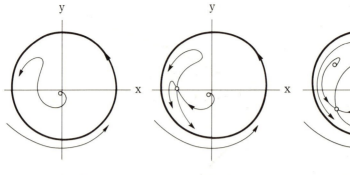
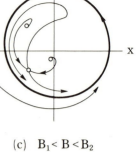
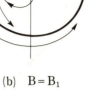

(a) $B < B_1$ (b) $B = B_1$ (c) $B_1 < B < B_2$

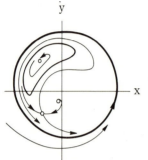
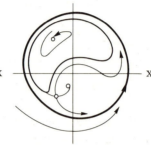
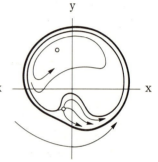

(d) $B_2 < B < B_3$ (e) $B_2 < B < B_3$ (f) $B = B_3$

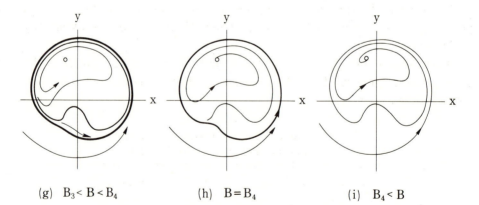

(g) $B_3 < B < B_4$ (h) $B = B_4$ (i) $B_4 < B$

Fig. 5.5 Various phase portraits of Eqs. (5.1) showing the transition of
singularities and limit cycles.

As previously in Sec. 4.3, we again introduce the mean-square amplitude of
the limit cycle defined by

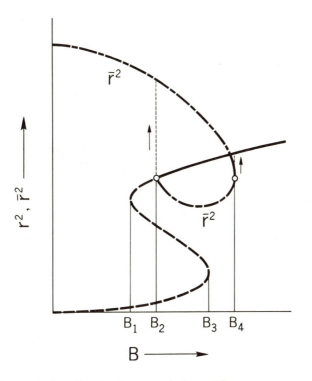

Fig. 5.6 Amplitude characteristics with mean-square
amplitude of the limit cycles.

$$\bar{r}^2 = \frac{1}{\tau} \oint_C r^2 dt \qquad (5.7)$$

where C is an integration path taken around the limit cycle and τ is the period
for the representative point $(x(t), y(t))$ to complete one revolution along the
limit cycle.

By using Eqs. (5.1) and (5.7), we have

$$\oint_C \left(\frac{\partial X}{\partial x} + \frac{\partial Y}{\partial y} \right) dt = \mu \int_0^\tau (1 - 2\gamma\nu^2 r^2) dt = 2\mu\tau\gamma\nu^2 \left(\frac{1}{2\gamma\nu^2} - \bar{r}^2 \right). \qquad (5.8)$$

From Poincaré's criterion for orbital stability [19], we see that a stable limit
cycle has mean-square amplitude greater than $1/2\gamma\nu^2$, an unstable limit cycle
less than $1/2\gamma\nu^2$, and a limit cycle of higher order has mean-square amplitude
equal to $1/2\gamma\nu^2$.

The results obtained above are summarized and sketched in Fig. 5.6. In
the figure the mean-square amplitude $\bar{r}^2$ of the limit cycle is plotted by uneven

dashed lines, and the arrows show jump phenomena between stable steady states when B is varied.* That is to say, in physical term, an entrained oscillation is sustained for $B_2 < B$, a beat oscillation for $B < B_4$, and hysteresis occurs between B_2 and B_4.

Thus far singularities and limit cycles of the autonomous system (5.1) have been investigated. Particular attention has been directed to the transitions of the phase portraits in the left hand region ($\nu < \omega_0$) on the $B\nu$ plane, and the example illustrating such transitions has been given. However, the application of these results to the original non-autonomous Rayleigh/Duffing mixed type equation requires further examination, since the representative point moves along the limit cycle rather quickly. It should be added that a region such as that represented by the curvilinear triangle BDG in Fig. 5.2 and in Fig. 4.2 does not appear in the self-oscillatory system having linear restoring force represented by the Eq. (3.1).

---

*When $B = 0$, Eqs. (5.1) are easily integrated; that is

$$r^2 = \frac{\omega_0^2 r_0^2}{(\omega_0^2 - \nu^2 r_0^2)e^{-\mu t} + \nu^2 r_0^2}$$

where

$$r_0^2 = r^2|_{t=0}.$$

We see that the limit cycle in this particular case is a circle centered at the origin with radius $r^2 = \omega_0^2/\nu^2 = 1/\gamma\nu^2$.

REFERENCES

1. C. Hayashi, Subharmonic oscillations in nonlinear systems, *J. Appl. Phys.* **24**, 521-529 (1953).

2. C. Hayashi, Initial conditions for certain types of nonlinear oscillations, *Proc. Symp. Nonlinear Circuit Analysis*, Vol. 6, pp. 63-92, Polytechnic Institute of Brooklyn, New York (1956).

3. C. Hayashi, Y. Nishikawa and M. Abe, Subharmonic oscillations of order one-half, *Trans. IRE Circuit Theory* **CT-7**, 102-111 (1960).

4. C. Hayashi and Y. Nishikawa, Initial conditions leading to different types of periodic solutions for Duffing's equation, *Proc. Symp. Nonlinear Oscillations (Intern. Union Theoret. Appl. Mech., Kiev)*, Vol. 2, pp. 377-393 (1963).

5. C. Hayashi, *Forced Oscillations in Nonlinear Systems*. Nippon Printing and Publishing Co., Osaka, Japan (1953).

6. G. Duffing, *Erzwungene Schwingungen bei veränderlicher Eigenfrequenz und ihre technische Bedeutung*. Vieweg, Braunschweig (1918).

7. C. Hayashi, *Nonlinear Oscillations in Physical Systems*. McGraw-Hill, New York (1964).

8. Y. Nishikawa, *A Contribution to the Theory of Nonlinear Oscillations*. Nippon Printing and Publishing Co., Osaka, Japan (1964).

9. L. S. Pontryagin, *Ordinary Differential Equations (in Russian)*. Fizmatgiz, Moscow (1961); English Translation, Pergamon Press, London (1962).

10. N. Minorsky, *Nonlinear Oscillations*. Van Nostrand, Princeton, NJ (1962).

11. G. Floquet, Sur les équations différentielles linéaires à coefficients périodiques, *Ann. Ecole Norm. Super.* **2-12**, 47-88 (1883).

12. G. W. Hill, On the Part of the Motion of the Lunar Perigee, *Acta Math.* **8**, 1-36 (1886).

13. N. W. McLachlan, *Theory and Application of Mathieu Functions*. Oxford University Press, London (1947).

14. E. T. Whittaker and G. N. Watson, *A Course of Modern Analysis*. Cambridge University Press, London (1935).

15. K. W. Blair and W. S. Loud, Periodic solution of $x'' + cx' + g(x) = Ef(t)$ under variation of certain parameters, *J. Soc. Ind. Appl. Math.* **8**, 74-101 (1960).

16. É. Mathieu, Mémoire sur le mouvement vibratoire d'une membrance de forme elliptique, *J. Math.* **2-13**, 137-203 (1868).

17. A. A. Andronow and A. A. Witt, Zur Theorie des Mitnehmens von van der Pol, *Arch. Elektrotech.* **24**, 99-110 (1930).

18. M. L. Cartwright, Forced oscillations in nearly sinusoidal systems, *J. Inst. Elec. Engrs (London)* **95**, 88-96 (1948).

19. J. J. Stoker, *Nonlinear Vibrations.* Interscience Publishers, New York (1950).

20. W. J. Cunningham, *Introduction to Nonlinear Analysis.* McGraw-Hill, New York (1958).

21. C. Hayashi, H. Shibayama and Y. Nishikawa, Frequency entrainment in a self-oscillatory system with external force, *Trans. IRE Circuit Theory* **CT-7**, 413-422 (1960).

22. B. R. Nag, Ultraharmonic and subharmonic resonance in an oscillator, *J. Brit. IRE* **19**, 411-416 (1959).

23. P. Riasin, Einstellungs- und Schwebungsprozesse bei der Mitnahme, *J. Tech. Phys. (USSR)* **2**, 195-214 (1935).

24. C. Hayashi, H. Shibayama and Y. Ueda, Quasi-periodic oscillations in a self-oscillatory system with external force, *Proc. Symp. Nonlinear Oscillations (Intern. Union Theoret. Appl. Mech., Kiev)*, Vol. 1, pp. 495-509 (1963).

25. B. van der Pol, On "Relaxation Oscillations", *Phil. Mag.* **7-2**, 978-992 (1926).

26. B. van der Pol, Forced oscillations in a circuit with nonlinear resistance, *Phil. Mag.* **7-3**, 65-80 (1927).

27. N. Kryloff and N. Bogoliuboff, Introduction to nonlinear mechanics, *Ann. Math. Studies*, No. 11, Princeton University Press, Princeton, NJ (1947).

28. A. A. Andronov and S. E. Chaikin, *Theory of Oscillations.* Princeton University Press, Princeton, NJ (1949); enlarged 2d ed. by A. A. Andronov, A. A. Witt and S. E. Chaikin, Fizmatgiz, Moscow (1959).

29. N. N. Bogoliuboff and Yu. A. Mitropolsky, *Asymptotic Methods in the Theory of Nonlinear Oscillations (in Russian).* Fizmatgiz, Moscow (1963); English translation, Gordon and Breach, Science Publishers, New York (1961).

30. A. Hurwitz, Über die Bedingungen, unter Welchen eine Gleichung nur Wurzeln mit negativen reellen Teilen besitzt, *Math. Ann.* **46**, 273-284 (1895).

31. H. Poincaré, *Oeuvres.* Vol. 1, Gauthier-Villars, Paris (1928).

32. E. A. Coddington and N. Levinson, *Theory of Ordinary Differential Equations.* McGraw-Hill, New York (1955).

33. I. Bendixon, Sur les courbes définies par des équations différentielles, *Acta Math.* **24**, 1-88 (1901).

34. M. L. Cartwright, Forced oscillations in nonlinear systems, *Ann. Math. Studies*, No. 20, pp. 149-241, Princeton University Press, Princeton, NJ (1950).

35. A. W. Gillies, On the transformations of singularities and limit cycles of the variational equations of van der Pol, *Quart. J. Mech. and Appl. Math.* (*London*) **7-2**, 152-167 (1954).

36. A. W. Gillies, The singularities and limit cycles of an autonomous system of differential equations of the second order associated with nonlinear oscillation, *Proc. Symp. Nonlinear Oscillations* (*Intern. Union Theoret. Appl. Mech., Kiev*), Vol. 2, pp. 134-155 (1963).

ACKNOWLEDGMENTS

The author owes a lasting debt of gratitude to Professor Dr. C. Hayashi, who has suggested the field of research of the present thesis and given him constant and generous guidance and encouragement in promoting this work.

In the preparation of the present paper the author was greatly aided by Professor H. Shibayama of the Osaka Institute of Technology, by Assistant Professor Y. Nishikawa and by Lecturer M. Abe both of Kyoto University who gave him valuable suggestions and much good advice of all kinds. Acknowledgment must also be made to Mr. S. Hiraoka and Mr. M. Kuramitsu for their excellent cooperation.

The KDC-I Digital Computer Laboratory of Kyoto University has made time available to the author. The author wishes to acknowledge the kind considerations of the staff of that organization. Finally, the author appreciates the assistance he received from Miss M. Takaoka, who typed the manuscript.

Selection 2

ON THE BEHAVIOR OF SELF-OSCILLATORY SYSTEMS WITH EXTERNAL FORCE

Chihiro Hayashi, Yoshisuke Ueda, Norio Akamatsu and Hidekiyo Itakura, Members

Department of Electrical Engineering, Kyoto University
(*Received* 9 *September* 1969)

Abstract

Nonlinear systems may exhibit oscillatory phenomena which are essentially different from those in linear systems. Among them there are nonlinear resonance with hysteresis, higher harmonic and subharmonic oscillations, and almost-periodic oscillations arising through self-excitation. This paper deals with nonlinear oscillations which occur in self-oscillatory systems under the action of a periodic external force by applying the mapping procedure based on the qualitative theory of differential equations.

1. INTRODUCTION

Let us consider the second-order nonlinear differential equation

$$\frac{d^2x}{dt^2} + f\left(x, \frac{dx}{dt}\right)\frac{dx}{dt} + g(x) = e(t) \tag{1}$$

where $e(t)$ is periodic function of period L. Oscillatory phenomena governed by this type of differential equation, i.e., nonlinear oscillations, received much attention through the oscillatory phenomena in mechanical and electrical systems first studied by Duffing and van der Pol. Since then, the phenomena have been studied by many engineers and mathematicians.

From both the engineering and mathematical points of view, the types and numbers of steady solutions become a first subject of discussion in relation to the form of non-linearity and the parameter values in the equation. Then the global behavior of solutions both in time and in phase space attracts our interest. In many cases, the relatively simple form of the equation is deceptive, and closer

75

study reveals complicated behavior. In this paper the steady solutions for a few equations of the form (1) are discussed by applying the method of harmonic balance and by using electronic computers. Further, global behavior of solutions is examined topologically by applying the transformation theory of differential equations.

2. METHOD OF ANALYSIS

In this section, the transformation theory for the topological analysis of global behavior of solutions is briefly explained.

2.1 Transformation T

Let us rewrite Eq. (1) as a system of first-order equations

$$
\begin{aligned}
\frac{dx}{dt} &= y & &\equiv X(x, y, t) \\
\frac{dy}{dt} &= -f(x, y)y - g(x) + e(t) &&\equiv Y(x, y, t)
\end{aligned}
\tag{2}
$$

where the functions f and g are differentiable functions of sufficient order with respect to x and y, and $e(t)$ is a periodic function of t with period L.

Let $\{x(x_0, y_0, t), y(x_0, y_0, t)\}$ be a solution of Eqs. (2) which when $t = 0$ is at the point $P_0(x_0, y_0)$ of the xy plane. Let us focus our attention on the location of the point $P_n(x_n, y_n)$ whose coordinates are given by

$$
x_n = x(x_0, y_0, nL), \qquad y_n = y(x_0, y_0, nL)
\tag{3}
$$

for $n = 0, \pm 1, \pm 2, \cdots$. Then the mapping

$$
P_n = T^n P_0
\tag{4}
$$

is defined which takes $P_0(x_0, y_0)$ into $P_n(x_n, y_n)$ of the xy plane. The mapping T thus defined is known to be a one-to-one, continuous and orientation-preserving transformation of the xy plane into itself [1, 2].

A point in the xy plane which is invariant under the mapping T is called a fixed point. Thus, if P_0 is a fixed point, then $P_0 = P_1 (= TP_0)$. Let m be the smallest positive integer for which $P_0 = P_m (= T^m P_0)$, then P_0 will iterate periodically through a set of m distinct points and is therefore called an m-periodic point. The set of these m-periodic points is called an m-periodic group. A solution of Eqs. (2) which passes through a fixed point of T at $t = 0$ is periodic with period L. Similarly, a solution associated with an m-periodic point is a subharmonic solution of period mL.

2.2 Maximum Finite Invariant Set

Under conditions generally met in practice [3], through any point in the xy plane sufficiently remote from the origin, there exist a simple closed curve having the property that every solution $(x(t), y(t))$ of Eqs. (2) which passes through the curve at time t can intersect the curve only by crossing it from the domain exterior to the curve into the domain interior to the curve with increasing t. Let Γ_0 denote a simple closed curve of this type, $T^n\Gamma_0$ be denoted by Γ_n and Δ_n be the closed domain bounded by Γ_n in the xy plane; then $\Delta_{n+1} \subset \Delta_n$ follows from the property of the curve Γ_0. The closed set defined by the intersection of all $\Delta_n (n \geq 0)$ is called the maximum finite invariant set for Eqs. (2). Let this set be denoted by Δ. It is known that Δ is a bounded and connected closed set and possesses the property that images of a point which is not contained in Δ tend to the maximum finite invariant set under iteration of the mapping T. Moreover, fixed points and periodic points, if they exist, are all contained in Δ. There exists at least one fixed point in Δ [1, 2].

2.3 Fixed Points, Periodic Points and Related Properties

In connection with the type and the number of periodic solutions of Eqs. (2), let us describe the classification of fixed points and periodic points and the number of these points contained in the maximum finite invariant set Δ.

Inasmuch as any neighboring point of the fixed point P_0 is taken into a neighborhood of the point P_0 under T, the type of fixed points can be classified by investigating the movement of neighboring images under the mapping T. Let a neighboring point $Q_0(x_0 + u_0, y_0 + v_0)$ of the fixed point $P_0(x_0, y_0)$ be transformed into $Q_1(x_0 + u_1, y_0 + v_1)$ under the mapping T, then the following relation results,

$$\begin{aligned} x_0 + u_1 &= x(x_0 + u_0, y_0 + v_0, L) \\ y_0 + v_1 &= y(x_0 + u_0, y_0 + v_0, L). \end{aligned} \tag{5}$$

Since P_0 is a fixed point under T, $u_0 = v_0 = 0$ means $u_1 = v_1 = 0$. For small values of u_0 and v_0, u_1 and v_1 can be expanded into power series in u_0 and v_0 as follows,

$$\begin{aligned} u_1 &= au_0 + bv_0 + \cdots \\ v_1 &= cu_0 + dv_0 + \cdots \end{aligned} \tag{6}$$

where

$$a = \frac{\partial x(x_0, y_0, L)}{\partial x_0}, \qquad b = \frac{\partial x(x_0, y_0, L)}{\partial y_0}$$
$$c = \frac{\partial y(x_0, y_0, L)}{\partial x_0}, \qquad d = \frac{\partial y(x_0, y_0, L)}{\partial y_0}. \tag{7}$$

The terms not explicitly given in the right sides of Eqs. (6) are of degree higher than the first in u_0 and v_0. Equations (6) express the mapping $(u_0, v_0) \rightarrow (u_1, v_1)$ in the neighborhood of the fixed point P_0, and this transformation is characterized by the roots m_1 and m_2 of the characteristic equation*

$$\begin{vmatrix} a - m & b \\ c & d - m \end{vmatrix} = 0. \tag{8}$$

The fixed point P_0 is called simple if both $|m_1|$ and $|m_2|$ are different from unity. Simple fixed points are classified as follows [1]:

Completely stable if	$|m_1| < 1, \quad |m_2| < 1$
Completely unstable if	$|m_1| > 1, \quad |m_2| > 1$
Directly unstable if	$0 < m_1 < 1 < m_2$
Inversely unstable if	$m_1 < -1 < m_2 < 0$.

The same classification also applies to periodic points.

Images of any point in the neighborhood of a completely stable fixed point tend to the fixed point under iteration of the mapping T. In the completely unstable case images move away from the fixed point. In the directly and inversely unstable cases there exist two invariant curves which traverse the fixed point, and successive images of the mapping T approach the fixed point along one of the invariant curves, while they move away from the fixed point along the other invariant curve. The former invariant curve is called the ω-branch and the latter the α-branch.

N. Levinson [1] and J. L. Massera [4] have discussed the number of fixed points and periodic points of Eqs. (2). Let $N(n)$ be the total number of n-periodic points (fixed points for $n = 1$) and $C(n)$ the total number of completely stable and completely unstable n-periodic points. Similarly, let $D(n)$ and $I(n)$ be the number of directly and inversely unstable n-periodic points, respectively.

---

*The product of the roots of Eq. (8) is given by

$$m_1 m_2 = \begin{vmatrix} a & b \\ c & d \end{vmatrix} = exp \int_0^L \left(\frac{\partial X}{\partial x} + \frac{\partial Y}{\partial y} \right) dt > 0.$$

If Eqs. (2) have a maximum finite invariant set and all periodic points are simple, the following relations hold.

For $n = 1$,

$$C(1) + I(1) = D(1) + 1, \qquad N(1) = 2D(1) + 1$$

For $n = 2, 4, 6, \cdots$,

$$C(n) + I(n) = D(n) + 2I(n/2), \qquad N(n) = 2[D(n) + I(n/2)] \qquad (9)$$

For $n = 3, 5, 7, \cdots$,

$$C(n) + I(n) = D(n), \qquad N(n) = 2D(n).$$

2.4 Invariant Closed Curves and Almost Periodic Solutions

A closed curve invariant under the mapping T is called an invariant closed curve. Clearly, if such curves exits they are contained in the maximum finite invariant set Δ. Now consider a solution of Eqs. (2) correlated with the invariant closed curve C. The solutions of Eqs. (2) which start from C at $t = 0$ form a surface S in the xyt space. Since C is invariant under T, the part of this surface lying between $t = 0$ and $t = L$ can be mapped onto a closed torus. Therefore the solutions of Eqs. (2) emanating from C can be investigated by the following differential equation on a torus [1]

$$\frac{d\theta}{dt} = p(\theta, t) \qquad (10)$$

where p is periodic in θ with period 1 and in t with period L. This type of differential equation has been studied by H. Poincaré [5], A. Denjoy [6] and P. Bohl [7]. Here let us briefly describe their results concerning the transformation of C into itself [8].

Associated with the solution curves of the equation (10) on the torus is a rotation number ρ. This number is the average advance of θ for an advance of t by L. If ρ is rational and of the form p/q, where p and q have no common factors, then there exist some periodic groups (which consist of q-periodic points) on C. The images of any point on C approach one of these periodic groups under iteration of the mapping T. In this case the solutions of Eqs. (2) include subharmonics of order q. If ρ is irrational there are two possibilities. One of them is termed the ergodic or transitive case. This is the case in which the derived set, or closure, of the sequence $\{P_i\}$ ($P_i = T^i P_0$, for $i = 0, \pm 1, \pm 2, \cdots$, where P_0 is an arbitrary point on C) coincides with C itself. The corresponding solution of Eqs. (2) is known as an almost periodic function of time t defined by H. Bohr

[9]. The other possibility is termed the singular or intransitive case. Here the derived set of $\{P_i\}$ is nowhere dense on C.

2.5 Doubly Asymptotic Points

As previously stated there exist invariant α- and ω-branches for a directly or inversely unstable fixed or periodic point. Here let us explain the behavior of these branches studied by Poincaré [10]. Let us consider the totality of α- and ω-branches of periodic points of all orders in the xy plane. It is readily seen from the uniqueness of solutions of differential equations that no α- (or ω-) branch can intersect another α- (or ω-) branch. However, an α-branch may intersect an ω-branch, and images of a point of intersection are themselves points of intersection, which converge along the ω-branch toward the unstable point on indefinite iteration of T, and along the α-branch on iteration of T^{-1}. Based on this property, the points of intersection are called doubly asymptotic points. A doubly asymptotic point is said to be of general type if the α- and ω-branches are not coincident or merely tangent at the doubly asymptotic point; in the contrary case the point is of special type. A doubly asymptotic point is called homoclinic if the α- and ω-branches on which it lies issue from the same point or from two points belonging to the same periodic group. A homoclinic point of the former type is called simple. A doubly asymptotic point is called heteroclinic if the α- and ω-branches on which it lies issue from two fixed and/or periodic points, each of them belonging to different periodic groups.

3. ENTRAINMENT OF FREQUENCY AND BEAT OSCILLATIONS

By applying transformation theory described above, here let us consider oscillatory phenomena which occur in periodically-forced self-oscillatory systems composed of a negative-resistance element or vacuum-tube with a feedback circuit. When a periodic force is applied to a self-oscillatory system, the frequency of the self-exited oscillation, that is, the natural frequency of the system, either falls in synchronism with the driving frequency, or else it fails to synchronize, resulting in the occurrence of a beat oscillation. The former phenomenon is known as synchronization or entrainment of frequency. Regions of the external force giving rise to frequency entrainment and global phase-plane structures representing oscillatory phenomena are investigated below for some forced self-oscillatory systems taking the form of Eq. (1).

3.1 Entrainment of Frequency

As a specific example of Eq. (1) let us consider the differential equation (see Appendix I)

$$\frac{d^2x}{dt^2} - \mu(1 - \gamma x^2)\frac{dx}{dt} + x^3 = B\cos\nu t, \quad \mu > 0. \tag{11}$$

The regions of frequency entrainment for this equation will be obtained by applying the method of harmonic balance and by using an analog computer. Further, phase-plane analysis is carried out by applying the transformation theory.

(a) Regions of Frequency Entrainment

To begin with let us consider the self-excited oscillation for Eq. (11) with $B = 0$, that is,

$$\frac{d^2x}{dt^2} - \mu(1 - \gamma x^2)\frac{dx}{dt} + x^3 = 0. \tag{12}$$

Let us express approximately the self-excited oscillation by the form

$$x(t) = a_0\cos\omega_0 t. \tag{13}$$

Substituting Eq. (13) into Eq. (12) and equating the coefficients of the terms containing $\sin\omega_0 t$ and $\cos\omega_0 t$ separately to zero, the following relations are obtained for the amplitude a_0 and the frequency ω_0,

$$a_0 = \sqrt{4/\gamma}, \qquad \omega_0 = \sqrt{3/\gamma}. \tag{14}$$

When the external force is applied and its frequency ν is nearly equal to ω_0, an approximate solution of the following form

$$x(t) = b_1(t)\sin\nu t + b_2(t)\cos\nu t \tag{15}$$

is introduced for the Eq. (11), where the functions $b_1(t)$ and $b_2(t)$ are constants for the entrained states; they are slowly varying functions of t when the system is not entrained. Substituting Eq. (15) into Eq. (11) and equating the coefficients of the terms containing $\cos\nu t$ and $\sin\nu t$ separately to zero leads to

$$\begin{aligned}
\frac{dx_1}{dt} &= \frac{\mu}{2}\left[(1 - r_1^2)x_1 - \sigma y_1 + \frac{B}{\mu\nu a_0}\right] \\
\frac{dy_1}{dt} &= \frac{\mu}{2}\left[\sigma x_1 + (1 - r_1^2)y_1\right]
\end{aligned} \tag{16}$$

where

$$x_1 = \frac{b_1}{a_0}, \quad y_1 = \frac{b_2}{a_0}, \quad r_1^2 = x_1^2 + y_1^2, \quad \sigma = \frac{\omega_0^2 r_1^2 - \nu^2}{\mu\nu}. \tag{17}$$

In deriving the autonomous equations (16), the following assumptions are used:

1. The amplitudes $b_1(t)$ and $b_2(t)$ are slowly varying functions of t; therefore, $d^2 b_1/dt^2$ and $d^2 b_2/dt^2$ are neglected.
2. Since μ is a small quantity, $\mu db_1/dt$ and $\mu db_2/dt$ are also discarded.

When the system is entrained, the amplitudes x_1 and y_1 become constant. The constants are obtained by putting $dx_1/dt = 0$ and $dy_1/dt = 0$, and a singular point representing an equilibrium state is given by

$$x_1 = -\frac{\mu\nu a_0}{B}(1 - r_1^2)r_1^2, \quad y_1 = \frac{\mu\nu a_0}{B}\sigma r_1^2 \tag{18}$$

where the amplitude r_1^2 is determined by the relation

$$[(1 - r_1^2)^2 + \sigma^2]r_1^2 = \left(\frac{B}{\mu\nu a_0}\right)^2. \tag{19}$$

The periodic solution thus obtained is actually sustained only when it is stable. In order to discuss the stability of the periodic solution, let us consider a small variation (ξ, η) from the singular point given by Eqs. (18). The variational equation is written as

$$\frac{d\xi}{dt} = a_1\xi + a_2\eta, \quad \frac{d\eta}{dt} = b_1\xi + b_2\eta \tag{20}$$

where

$$a_1 = \frac{\mu}{2}\left(1 - r_1^2 - 2x_1^2 - 2\frac{\omega_0^2}{\mu\nu}x_1 y_1\right), \quad a_2 = \frac{\mu}{2}\left(-2x_1 y_1 - \sigma - 2\frac{\omega_0^2}{\mu\nu}y_1^2\right)$$

$$b_1 = \frac{\mu}{2}\left(-2x_1 y_1 + \sigma + 2\frac{\omega_0^2}{\mu\nu}x_1^2\right), \quad b_2 = \frac{\mu}{2}\left(1 - r_1^2 - 2y_1^2 + 2\frac{\omega_0^2}{\mu\nu}x_1 y_1\right). \tag{21}$$

If the real parts of the two roots of the characteristic equation

$$\begin{vmatrix} a_1 - \lambda & a_2 \\ b_1 & b_2 - \lambda \end{vmatrix} = 0 \tag{22}$$

are negative, the corresponding equilibrium state is stable. This stability con-

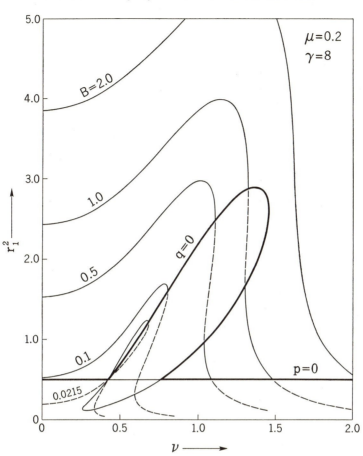

Fig. 1 Resonance curves for the approximate harmonic solution of Eq. (11).

dition is given by the Routh-Hurwitz criterion, namely,

$$p = -(a_1 + b_2) = \mu(2r_1^2 - 1) > 0$$

$$q = a_1 b_2 - a_2 b_1 = \frac{\mu^2}{4}\left[(1 - r_1^2)(1 - 3r_1^2) + \sigma^2 + 2\frac{\omega_0^2}{\mu\nu}\sigma r_1^2\right] > 0. \tag{23}$$

Figure 1 shows the resonance curves obtained by plotting Eq. (19) in the (ν, r_1^2) plane for several values of the amplitude B. The system parameters in Eq. (11) are given by $\mu = 0.2$ and $\gamma = 8$. The stability limits are given by $p = 0$ and $q = 0$ of Eqs. (23) and they are also shown in the figure. Hence the solid lines of the resonance curves represent stable states, while the dashed portions indicate

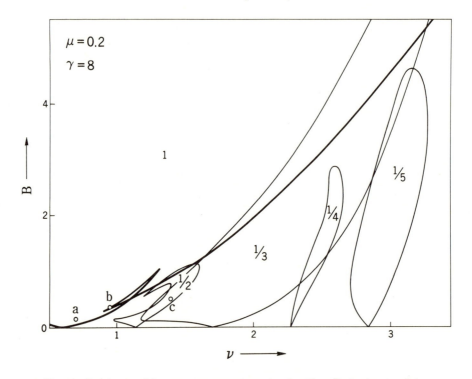

Fig. 2 Regions of frequency entrainment for Eq. (11) obtained by
analog computer analysis.

unstable states. When the driving frequency ν is nearly equal to the natural
frequency $\omega_0\,(= 0.61\cdots)$, the amplitude of the oscillation becomes large even
for small B owing to the resonance between these frequencies. As the driving
frequency ν leaves the natural frequency ω_0, the amplitude becomes small.
Thus once the response curves of the entrained oscillations are obtained, the
approximate region of harmonic entrainment can be obtained in the $B\nu$ plane,
as the region in which Eqs. (16) has at least one stable equilibrium point.[†]
In order to check these results, solutions of Eq. (11) are sought by using an
analog computer. The region of harmonic entrainment, and some of the main
regions of subharmonic entrainment, are shown in Fig. 2. In the area common
to entrainment regions of different frequencies, final oscillations are determined
depending on the initial conditions. If the amplitude B and the frequency ν
are given outside these regions, non-periodic beat oscillations occur.

[†]The foregoing analysis of the entrained oscillation is based on the autonomous equations
(16); phase-portraits for Eqs. (16) showing generation and extinction of the singular points
caused by changes of the external force are reported in Refs. [11] and [12].

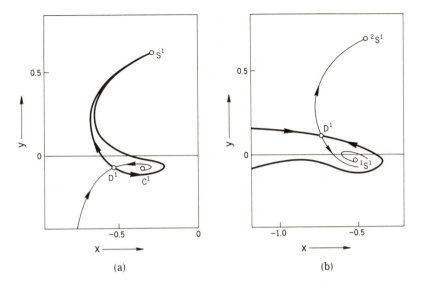

Fig. 3 Fixed points and invariant curves representing solutions of
Eqs. (24), the parameters being
(a) $B = 0.1$, $\nu = 0.7$
(b) $B = 0.36$, $\nu = 0.95$.

(b) Phase-Plane Analysis by Applying Mapping Method
    Equation (11) is transformed into a set of simultaneous equations

$$\frac{dx}{dt} = y, \qquad \frac{dy}{dt} = 0.2(1 - 8x^2)y - x^3 + B \cos \nu t. \tag{24}$$

By applying the mapping method described in Sec. 2, the topological properties
of the solutions of Eqs. (24) are examined by using computers.
    (i) Harmonic Entrainment   Let us consider the solutions of Eqs. (24) in
which the parameters are $B = 0.1$ and $\nu = 0.7$. The location of the parameters
is chosen in the region of harmonic entrainment of Fig. 2 (indicated by point
a). Figure 3(a) shows the phase-plane portrait for this case. In the figure
point D^1 is a directly unstable fixed point; S^1 and C^1 indicate completely
stable and completely unstable fixed points, respectively. Besides these fixed
points the invariant curves which abut D^1 are also shown. The arrows on the
invariant curves indicate the direction of the movement of successive images
under the mapping T. The two α-branches (heavy line) which start from D^1
both terminate at the point S^1; the closed region bounded by these α-branches
is the maximum finite invariant set Δ. Next let us consider a case in which the
external force is chosen in the region where two types of harmonic entrainments

85

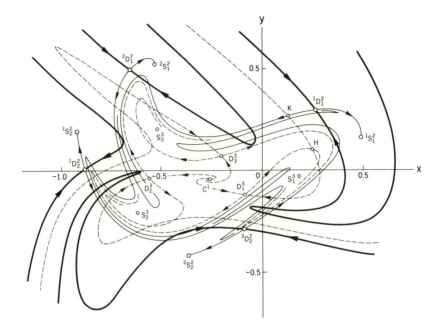

Fig. 4 Fixed point, periodic points, and invariant curves of the mapping
for Eqs. (24), the parameters being
$B = 0.5$ and $\nu = 1.4$.

coexist. The parameters are $B = 0.36$ and $\nu = 0.95$ (point b in Fig. 2). The phase-plane portrait is shown in Fig. 3(b). Point D^1 is a directly unstable fixed point and points $^1S^1$ and $^2S^1$ are both completely stable fixed points corresponding to the non-resonant and resonant states, respectively. The ω-branches (heavy line) are the boundaries of the two domains of attraction which contain fixed points $^1S^1$ and $^2S^1$.

(ii) 1/2-Harmonic and 1/3-Harmonic Entrainments   Figure 4 shows the phase-plane portrait for the case in which the external force lies in the area common to the regions of 1/2- and 1/3-harmonic entrainments. The parameters of Eqs. (24) are $B = 0.5$ and $\nu = 1.4$ (indicated by point c in Fig. 2). In this figure the symbol $^iD^n_j$ denotes a directly unstable n-periodic point which belongs to the i-th periodic group and the subscript j shows the order of movement of images under the mapping T, that is,

$$T(^iD^n_j) = {}^iD^n_{j+1}, \qquad T^n(^iD^n_j) = {}^iD^n_j. \tag{25}$$

Similar notations are used for completely stable periodic points (S), completely unstable periodic points (C), and inversely unstable periodic points (I). Invariant curves of the mapping T^2 are shown by solid lines and those of the mapping T^3 by dashed lines. It is seen in the figure that the α- and ω-branches of the directly unstable periodic points intersect one another and that there exist some doubly asymptotic points. For example, since the point H is an intersection of the α-branch and ω-branch both emanating from D_1^3, this point is a simple homoclinic point. The images of the point H converge toward D_1^3 on indefinite iteration of both T^3 and T^{-3}. Also since K is an intersection of the α-branch emanating from $^1D_1^2$ and the ω-branch converging to D_1^3, this point is a heteroclinic point. The images of the point K converge toward D_1^3 on indefinite iteration of T^3 and toward $^1D_1^2$ under T^{-2}. The points H and K are examples of doubly asymptotic points. It can be seen from the figure that there must exist an infinite number of doubly asymptotic points.[‡] In this case the maximum finite invariant set and the domains of attraction for different completely stable periodic groups exhibit extremely complicated configurations and they cannot be drawn in the straightforward manner of Fig. 3.

3.2 Analysis of Beat Oscillations

Let us consider beat oscillations which occur in a system described by the differential equation

$$\frac{d^2x}{dt^2} - \mu\left[1 - \gamma\left(\frac{dx}{dt}\right)^2\right]\frac{dx}{dt} + x^3 = B\cos\nu t, \quad \mu > 0. \qquad (26)$$

Following the same procedure as before, the resonance curves are drawn for approximate solutions which consist only of the fundamental component. The result is shown in Fig. 5. The system parameters are given by $\mu = 0.2$ and $\gamma = 4$ in Eq. (26). In the figure the amplitude r_1^2 shows a normalized quantity in the same way as in Fig. 1 (see Appendix II). From this result we can obtain the approximate region of harmonic entrainment. Then the solutions of Eq. (26) are examined by using an analog computer. In the course of this study, when the external force is given just outside the region of harmonic entrainment, different types of beat oscillations are observed depending on the magnitude of the amplitude of the external force. Let us examine the aspect of these solutions in the phase plane by applying the transformation theory.

[‡]The following theorems have been proved by G. D. Birkhoff [13-15]:

(a) An arbitrary neighborhood of a homoclinic point contains an infinite number of homoclinic points.

(b) An arbitrary neighborhood of a homoclinic point contains an infinite number of periodic points.

Some examples of periodic points are shown in Refs. [16-18].

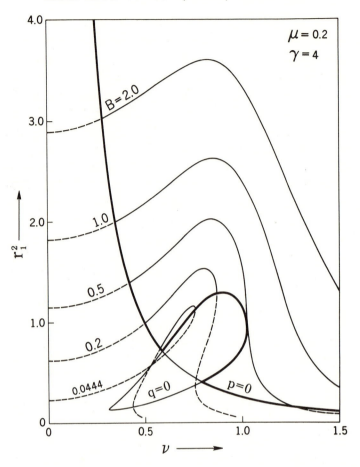

Fig. 5 Resonance curves for the approximate harmonic solution of
Eq. (26), the system parameters being $\mu = 0.2$ and $\gamma = 4$.

First let us consider the case in which the amplitude B of the external force is comparatively small. Figure 6(a) shows the successive movement of images obtained by using an analog computer. The system under consideration is described by

$$\frac{dx}{dt} = y, \qquad \frac{dy}{dt} = 0.2(1 - 4y^2)y - x^3 + B\cos\nu t \qquad (27)$$

with $B = 0.1$ and $\nu = 1.1$. In this figure the numbers attached to the points indicate the order of successive images under the mapping T, and these are counted after the transient has decayed. It is seen in the figure that these images form a smooth invariant closed curve, which is the boundary of the

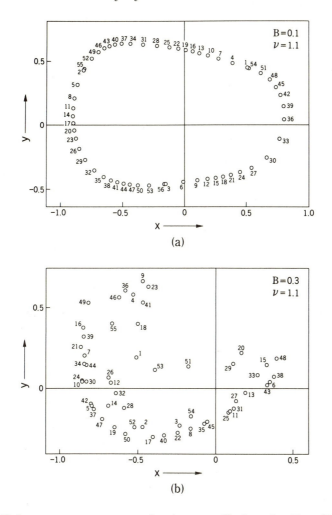

Fig. 6 Point sequences representing beat oscillations for Eqs. (27), the parameters being (a) $B = 0.1$, $\nu = 1.1$ and (b) $B = 0.3$, $\nu = 1.1$.

maximum finite invariant set in this case. The rotation number ρ associated with this curve is approximately equal to 0.66. It cannot be concluded by computer experiment whether it is a rational or an irrational number. If we assume that ρ is an irrational number, then the mapping is either ergodic or singular. However, it is known that the singular case cannot occur when an invariant closed curve is sufficiently smooth, as is the case in the figure [8, 19]. The mapping is thus ergodic and the solution of Eqs. (27) is almost periodic.

Next let us consider the case in which the amplitude B is increased. Figure 6(b) shows the behavior of successive images in the system (27) with $B = $

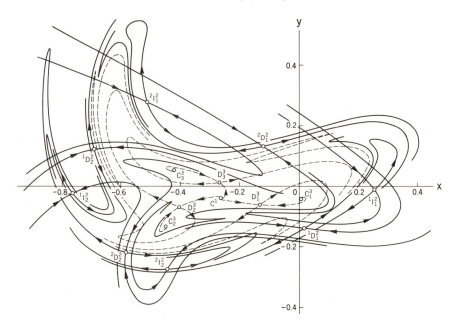

Fig. 7 Fixed point, periodic points and invariant curves of the mapping
for Eqs. (27), the parameters being $B = 0.3$ and $\nu = 1.1$.

0.3 and $\nu = 1.1$. The numbers attached to the images are counted after the transient has decayed in the same way as in Fig. 6(a). The movement of the images is irregular and complicated as is seen in the figure. In this case we cannot infer the existence of a simple invariant closed curve as we did in Fig. 6(a). Figure 7 shows the phase-plane portrait of this case. The same notation is applied to the fixed and periodic points, and solid and dashed lines are used as in Fig. 4. Again the existence of homoclinic and heteroclinic points is observed.

Let us consider the number of fixed and periodic points, using the same notation as in Sec. 2.3. Since $C(1) = 1, D(2) = I(2) = 4$ and $D(3) = C(3) = 3$, the relations (9) for $n = 1, 2$ and 3 are satisfied. For $n = 4$, $C(4) + I(4) = D(4) + 2I(2) \neq 0$. Computer analysis revealed that there are no completely stable or unstable 4-periodic points. That is, $C(4) = 0$, and therefore $I(4) \neq 0$ must hold. Similarly, $I(8) \neq 0, \cdots$ hold and the existence of 2^n-periodic points (n: an integer) is inferred.

Let us draw a simple closed curve Γ_0 in the xy plane as shown in Fig. 8 which possesses the property mentioned in Sec. 2.2. Then successive mapping is applied to Γ_0. For simplicity, only the curve Γ_3 $(= T^3\Gamma_0)$ is shown in the figure. From this result it can be seen that images Γ_i become more and more complicated as i increases, and the configuration of Γ_i for $i \to \infty$ cannot even

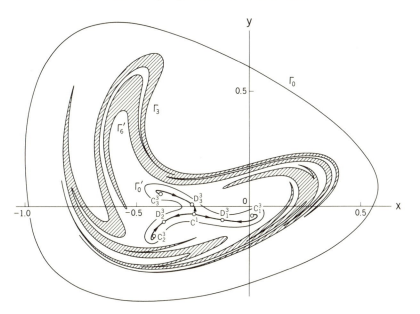

Fig. 8 Ring domain within which the images under the mapping representing
steady state irregular beat oscillations are confined.

be imagined. Next let us draw the closed curve Γ'_0 which encloses the points
C^1, C^3_i, D^3_i ($i = 1, 2, 3$). Let us consider the closed domain S_0 bounded by
Γ'_0 and apply successive mapping on S_0 to obtain the closed set $\bigcup^6_{i=0} T^i S_0$.
However, the configuration of this boundary curve is much too intricate, and
we are compelled to draw the closed curve Γ'_6 inside this closed domain in
somewhat simplified form. It is seen from this construction that successive
images P_i, which represent steady oscillation, must remain in the shaded ring
domain bordered by the curves Γ_3 and Γ'_6. The full implications of this point
sequence $\{P_i\}$ have not yet been clarified.

4. CONCLUSION

Nonlinear oscillatory phenomena have been studied which occur in periodically-forced self-oscillatory systems containing a negative-resistance element or
vacuum-tube with feedback circuit.

First, the resonance curve of the harmonic entrainment was discussed by the
method of harmonic balance; then the regions of entrainment were obtained at
fundamental and some principal subharmonic frequencies by using an analog

computer. From these results, shapes and mutual relations of the regions at various frequencies are clarified.

Second, in order to understand the oscillatory phenomena more clearly, some representative phase-plane portraits were obtained by applying the transformation theory based on the qualitative theory of differential equations. For certain values of the external force, the existence of doubly asymptotic points was observed in the phase plane. In this case, the maximum finite invariant set and the domains of attractions for completely stable fixed or periodic points exhibit complicated configurations. Further, for the external force prescribed outside the regions of entrainment, different types of beat oscillations were observed. One of them is a almost-periodic oscillation and the corresponding successive images form a smooth invariant closed curve on the phase plane as shown in Fig. 6(a). The other type of beat oscillations occurs depending upon the external force. The genesis of the oscillations is not fully grasped. In the neighborhood of the point sequence in the phase plane representing steady states of this type of oscillations there exist unstable 2^n-periodic points as well as doubly asymptotic points as shown in Fig. 7. Further, since all points of this sequence are found inside the ring domain of Fig. 8 and there are no stable periodic points, the resulting oscillation behaves non-periodically. The nature of this type of oscillation deserves attention as material for further study from the point of view of topological dynamics [20].

REFERENCES

1. N. Levinson, Transformation theory of non-linear differential equations of the second order, *Ann. Math.* **45**, 723-737 (1944).

2. S. Furuya, *Nonlinear Problems – Theory of Forced Oscillations* (*in Japanese*). Gendai Sugaku Kohza 12-C, Kyoritsu Shuppan, Tokyo (1957).

3. N. Levinson, On the existence of periodic solutions for second order differential equations with a forcing term, *J. Math. Phys.* **22**, 41-48 (1943).

4. J. L. Massera, The number of subharmonic solutions of non-linear differential equations of the second order, *Ann. Math.* **50**, 118-126 (1949).

5. H. Poincaré, Sur les courbes définies par les équations différentielles, *J. Math.* **4-1**, 167-244 (1885).

6. A. Denjoy, Sur les courbes définies par les équations différentielles à la surface du tore, *J. Math.* **11**, 333-375 (1932).

7. P. Bohl, Über die hinsichtlich der unabhängigen und abhängigen variabeln periodische differentialgleichung erster ordnung, *Acta Math.* **40**, 321-336 (1916).

8. E. A. Coddington and N. Levinson, *Theory of Ordinary Differential Equations.* McGraw-Hill, New York (1955).

9. H. Bohr, Zur theorie der fastperiodischen funktionen, *Acta Math.* **45**, 29-127 (1924); **46**, 101-214 (1925); **47**, 237-281 (1926).

10. H. Poincaré, *Les Méthodes Nouvelles de la Mécanique Céleste.* Vol. 3, Chap. 33, Gauthier-Villars, Paris (1899).

11. C. Hayashi, *Nonlinear Oscillations in Physical Systems.* McGraw-Hill, New York (1964).

12. C. Hayashi, M. Abe and Y. Ueda, Self-oscillatory system with nonlinear restoring force (in Japanese), *IECE Technical Report*, Nonlinear Theory (Dec. 8, 1962).

13. G. D. Birkhoff, On the periodic motions of dynamical systems, *Acta Math.* **50**, 359-379 (1927).

14. G. D. Birkhoff and P. A. Smith, Structure analysis of surface transformations, *J. Math.* (Liouville), S. 9, Vol. 7, pp. 345-379 (1928).

15. G. D. Birkhoff, Nouvelles recherches sur les systémes dynamiques, *Mem. Pont. Acad. Sci. Novi Lyncaei*, S. 3, Vol. 1, pp. 85-214 (1935).

16. C. Hayashi, Y. Ueda and H. Kawakami, Transformation theory as applied to the solutions of non-linear differential equations of the second order, *Int. J. Non-Linear Mech.* **4**, 235-255 (1969).

17. C. Hayashi, Y. Ueda and H. Kawakami, Periodic solutions of Duffing's equation with reference to doubly asymptotic solutions, *Proc. Fifth Int. Conf. Nonlinear Oscillations, Kiev*, Vol. 2, pp. 507-521 (1969).

18. C. Hayashi, H. Kawakami and Y. Ueda, Doubly asymptotic solutions of Duffing's equation (in Japanese), *IECE Technical Report*, Nonlinear Theory (July 17, 1969).

19. E. R. van Kampen, The topological transformations of a simple closed curve into itself, *Amer. J. Math.* **57**, 142-152 (1935).

20. G. D. Birkhoff, Surface transformations and their dynamical applications, *Acta Math.* **43**, 1-119 (1920).

APPENDICES

I. Derivation of Eq. (11)

The schematic diagram in Fig. A(a) shows a forced self-oscillatory circuit which contains a negative-resistance element N. With the notation of the figure,

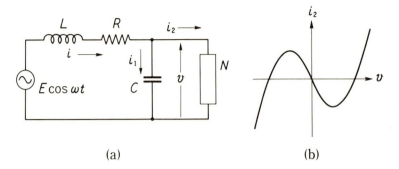

(a)   (b)

Fig. A (a) Negative-resistance oscillator with external force.
(b) Nonlinear characteristic of the negative-resistance element N.

the equations for the circuit are written as

$$L\frac{di}{dt} + Ri + v = E\cos\omega t, \quad i = i_1 + i_2, \quad i_1 = C\frac{dv}{dt}. \tag{A.1}$$

The nonlinear characteristic of the negative resistance N is shown in Fig. A(b) and is expressed by

$$i_2 = f(v). \tag{A.2}$$

Retaining v and eliminating the other variables, we obtain

$$LC\frac{d^2v}{dt^2} + \left(RC + L\frac{df}{dv}\right)\frac{dv}{dt} + v + Rf(v) = E\cos\omega t. \tag{A.3}$$

For simplicity, let the voltage-current characteristic be expressed by

$$i_2 = f(v) = Sv\left(-1 + \frac{v^2}{V_s^2}\right) \tag{A.4}$$

with the relation $S = 1/R$; then Eq. (A.3) is transformed into

$$\frac{d^2x}{d\tau^2} - \mu(1 - \gamma x^2)\frac{dx}{d\tau} + x^3 = B\cos\nu\tau \tag{A.5}$$

where

$$x = \frac{v}{V_s}, \quad \tau = \frac{t}{\sqrt{LC}}, \quad B = \frac{E}{V_s}, \quad \nu = \omega\sqrt{LC}$$

$$\mu = \frac{LS - RC}{\sqrt{LC}}, \qquad \gamma = \frac{3LS}{LS - RC}. \tag{A.6}$$

II. Approximate Harmonic Solution of Eq. (26) and Its Stability

Here let us give the approximate harmonic solution and its stability condition for the equation (26), i.e.,

$$\frac{d^2 x}{dt^2} - \mu \left[1 - \gamma \left(\frac{dx}{dt} \right)^2 \right] \frac{dx}{dt} + x^3 = B \cos \nu t, \quad \mu > 0. \tag{A.7}$$

Let the approximate solution of Eq. (A.7) with $B = 0$ be

$$x(t) = a_0 \cos \omega_0 t, \tag{A.8}$$

then the amplitude a_0 and the frequency ω_0 are given by

$$a_0 = \sqrt{\frac{4}{3\sqrt{\gamma}}}, \qquad \omega_0 = \sqrt{\frac{1}{\sqrt{\gamma}}}. \tag{A.9}$$

When $B \neq 0$, let us assume an approximate harmonic solution of the form

$$x(t) = b_1(t) \sin \nu t + b_2(t) \cos \nu t. \tag{A.10}$$

By a procedure similar to that in Sec. 3.1, a set of autonomous equations is derived, that is,

$$\frac{dx_1}{dt} = \frac{\mu}{2} \left[(1 - \sqrt{\gamma} \nu^2 r_1^2) x_1 - \sigma y_1 + \frac{B}{\mu \nu a_0} \right]$$
$$\frac{dy_1}{dt} = \frac{\mu}{2} \left[\sigma x_1 + (1 - \sqrt{\gamma} \nu^2 r_1^2) y_1 \right] \tag{A.11}$$

where

$$x_1 = \frac{b_1}{a_0}, \quad y_1 = \frac{b_2}{a_0}, \quad r_1^2 = x_1^2 + y_1^2, \quad \sigma = \frac{\omega_0^2 r_1^2 - \nu^2}{\mu \nu}. \tag{A.12}$$

Any singular point (for which x_1 and y_1 are constants) of Eqs. (A.11) is given by

$$x_1 = -\frac{\mu \nu a_0}{B} (1 - \sqrt{\gamma} \nu^2 r_1^2) r_1^2, \qquad y_1 = \frac{\mu \nu a_0}{B} \sigma r_1^2 \tag{A.13}$$

where r_1^2 is determined by solving the equation

$$[(1 - \sqrt{\gamma} \nu^2 r_1^2)^2 + \sigma^2] r_1^2 = \left(\frac{B}{\mu \nu a_0} \right)^2. \tag{A.14}$$

The stability condition of the singular point is given by

$$p = \mu(2\sqrt{\gamma}\nu^2 r_1^2 - 1) > 0$$

$$q = \frac{\mu^2}{4}\left[(1 - \sqrt{\gamma}\nu^2 r_1^2)(1 - 3\sqrt{\gamma}\nu^2 r_1^2) + \sigma^2 + 2\frac{\omega_0^2}{\mu\nu}\sigma r_1^2\right] > 0. \tag{A.15}$$

Figure 5 in Sec. 3.2 was plotted by making use of Eqs. (A.14) and (A.15).

Selection 3

COMPUTER SIMULATION OF NONLINEAR ORDINARY DIFFERENTIAL EQUATIONS AND NON-PERIODIC OSCILLATIONS

Yoshisuke Ueda*, Norio Akamatsu**
and Chihiro Hayashi*, Members

*Faculty of Engineering, Kyoto University
**Faculty of Engineering, Tokushima University
(*Received 7 September* 1972)

Abstract

There occur periodic and non-periodic oscillations in nonlinear oscillatory systems with periodic external force. These phenomena are examined by analyzing nonlinear differential equations describing the systems, i.e., the mathematical models of the phenomena. However, if every solution of the equations lacks the stability which would be associated with realizability in a physical system, how can the behavior of the corresponding oscillatory system be explained?

This paper discusses the relation between non-periodic oscillations in the computer-simulated systems and the exact solutions of second-order nonlinear ordinary differential equations of class D, that is, of dissipative systems for large displacements.

1. INTRODUCTION

In nonlinear oscillatory systems with periodic external force there sometimes occur steady non-periodic oscillations in addition to periodic oscillations whose fundamental frequency is the same as, or equal to a rational multiple of, the external frequency. These oscillatory phenomena are examined by analyzing differential equations, the mathematical models for the phenomena, by considering appropriate aspects of the phenomena. In particular, the evolution of the state as time progresses is studied by the behavior of the solutions of the differential equations, i.e., the movement of a representative point in the phase space. However, considering that, among the solutions of the differential

99

equation, only stable solutions can represent a realizable long-term oscillatory state in the actual physical system, how can the phenomenon be explained, and what kind of oscillations will be observed, in the system where every solution of the differential equation exhibits instability? When no analytical solutions of nonlinear differential equations can be expected, the problems presented here should be examined not only to clarify the phenomenon but also to assess the validity of numerical methods for the solution of the differential equations, because computer simulation techniques are spreading far and wide.

In this paper, non-periodic steady oscillations are observed in computer-simulated solutions of the second-order nonlinear ordinary differential equation

$$\frac{d^2x}{dt^2} + f\left(x, \frac{dx}{dt}\right)\frac{dx}{dt} + g(x) = e(t) \tag{1}$$

where $e(t)$ is a periodic function of the period L. Then the phenomena are described in terms of the appropriate mathematical concepts of recurrence.

2. OBSERVATION OF NON-PERIODIC OSCILLATIONS IN COMPUTER-SIMULATED SYSTEMS

As the differential equation (1) is periodic in t with the period L, the behavior of the system can effectively be analyzed by applying the transformation theory.* Let us denote by T the transformation which transforms the phase plane at $t = 0$ into itself at $t = L$ following the solution curves of the first-order system derived from Eq. (1) by the usual substitution $y = dx/dt$. Then, according to the definition of the transformation or the mapping T, the behavior of the solution passing through the point P at $t = 0$ in the phase plane is expressed by a complete sequence $\cdots, P_{-2}, P_{-1}, P, P_1, P_2, \cdots$ ($P_n = T^n P$, $n = 0, \pm 1, \pm 2, \cdots$) generated by the point P.

In this section, let us perform the computer simulation for the equations describing a forced oscillatory system and a forced self-oscillatory system as special cases of Eq. (1), and observe the movements of images P_n under the transformation T, and the global aspect of the solutions in the phase plane.

2.1 Non-periodic Oscillations in a Forced Oscillatory System

---

*For the concepts of the transformation theory used in this paper, see the Appendix and Refs. [2-10].

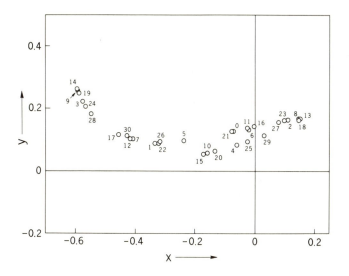

Fig. 1 Example of a point sequence representing steady oscillation
which occurs in the system described by Eqs. (3), the parameters
being $k = 0.2$, $B = 0.3$ and $B_0 = 0.08$.

Let us consider the Duffing equation

$$\frac{d^2 x}{dt^2} + k\frac{dx}{dt} + x^3 = B\cos t + B_0 \tag{2}$$

which describes a forced oscillatory system. This equation represents a mathematical model for a series-resonant circuit containing a saturable inductor under the impression of d.c. and sinusoidal voltage [1].

As is well known, the oscillation described by Eq. (2) may include resonant, non-resonant, higher-harmonic and subharmonic oscillations. However, whether Eq. (2) has non-periodic steady solutions has not been clarified, but the computer simulation result reveals the existence of a steady oscillation which must be considered non-periodic. In the following, let us give a few examples of such oscillations.

Figure 1 shows an example of the behavior of images P_n moving in the xy plane under the transformation T in a system described by the first-order simultaneous equations

$$\frac{dx}{dt} = y, \qquad \frac{dy}{dt} = -ky - x^3 + B\cos t + B_0 \tag{3}$$

with $k = 0.2, B = 0.3$ and $B_0 = 0.08$. Numerals attached to points in this

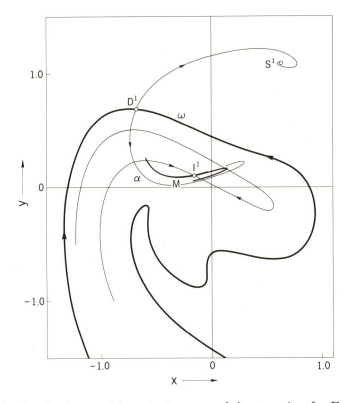

Fig. 2 Fixed points and invariant curves of the mapping for Eqs. (3), the parameters being $k = 0.2$, $B = 0.3$ and $B_0 = 0.08$.

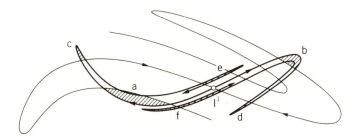

Fig. 3 Partial details of α- and ω-branches of the inversely unstable fixed point I^1.

figure indicate the order of the successive images under the transformation T. They are counted after the transient state has decayed. In every simulation, the pattern of this non-periodic oscillation in the phase plane (the shape of the point set representing the steady state) is reproducible, but the movement of

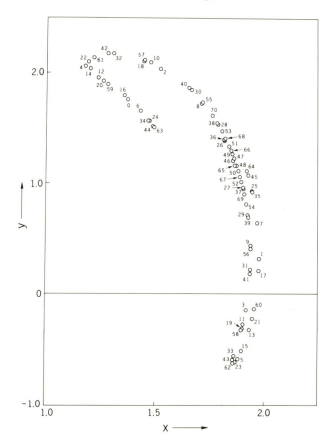

Fig. 4 Example of a point sequence representing steady oscillation
which occurs in the system described by Eqs. (3), the parameters
being $k = 0.2$, $B = 1.2$ and $B_0 = 0.85$.

the images cannot be reproduced over long intervals of time. This tendency is
particularly remarkable in simulation by analog computer. In digital simulation,
the movement of the images differs considerably depending on the integration
step size, and the schemes for numerical integration. Figure 2 shows the global
aspect of the fixed points and the invariant curves of the mapping in the xy
plane for this case. In the figure, points S^1, D^1 and I^1 are completely stable,
directly unstable and inversely unstable fixed points, respectively. The arrow on
the invariant curve indicates the direction of the movement of the images under
the transformation T. The ω-branches (the heavy solid line) of the directly
unstable fixed point divide the phase plane into two regions. When the initial
point is given inside the region containing S^1, the system settles down to the
periodic oscillation represented by the completely stable fixed point S^1. The

103

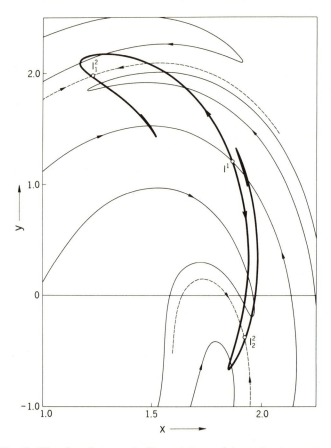

Fig. 5 Fixed point, periodic points and invariant curves of
the mapping for Eqs. (3), the parameters being
$k = 0.2$, $B = 1.2$ and $B_0 = 0.85$.

steady oscillation when the initial point is given inside the region containing I^1
is non-periodic as shown in Fig. 1. The image P_n representing the steady state
continues to move in the neighborhood of the α-branches (medium heavy solid
line) of the inversely unstable fixed point I^1.

Figure 3 shows the aspect of α-branches (heavy solid line) and ω-branches
(thin solid line) of the point I^1 schematically. Both branches intersect with each
other and form an infinite number of doubly asymptotic points. The α-branches
are asymptotic to themselves and are confined inside the bounded region, while
the ω-branches extend to infinity. The shaded region a in the figure is mapped
onto the regions b, c, d, $\cdots$ successively under the transformation T. From
these results, the configuration of α- and ω-branches of I^1 and the behavior of
the points on the branches under the transformation T can be understood.

Next, let us consider an example in which the external force takes a slightly larger amplitude. Figure 4 shows the simulation result of a point sequence representing the steady state of the images under the transformation T, defined by Eqs. (3) with $k = 0.2$, $B = 1.2$ and $B_0 = 0.85$. As in Fig. 1, the image traces out a pattern resembling a segment of curve. Figure 5 shows the fixed point, periodic points and invariant curves of the mapping for the corresponding parameter values. The heavy solid lines represent the α-branches, the thin solid lines the ω-branches of the inversely unstable fixed point I^1. As to the inversely unstable 2-periodic points I_1^2 and I_2^2, only the ω-branches are shown by the thin dashed lines. The images under the transformation T starting from any point in the phase plane continue to move, after a sufficiently long lapse of time, in the neighborhood of the α-branches of the inversely unstable fixed point I^1.

2.2 Non-periodic Oscillations in a Forced Self-Oscillatory System

Let us consider the equation of the form (1)

$$\frac{d^2x}{dt^2} - \mu(1 - x^2)\frac{dx}{dt} + x^3 = B\cos\nu t \tag{4}$$

which describes a self-oscillatory circuit containing a negative-resistance element with the injection of a sinusoidal signal [2].

When a periodic force is applied to a self-oscillatory system, the phenomenon of synchronization occurs in a certain band of the external frequency; the system exhibits harmonic, higher-harmonic or subharmonic oscillation having period the same as, an integral multiple or submultiple of, the driving frequency. If the amplitude and frequency of the external force do not permit such synchronization, then the oscillation in the system becomes non-periodic. The non-periodic oscillation is one of two types, almost periodic oscillation and others. The almost periodic oscillation is liable to occur when the amplitude of the external force is relatively small, and can be described by a completely stable invariant closed curve in the phase plane. In this section, let us observe the other type of non-periodic oscillations in computer-simulated systems.

Figure 6 shows an example of a point sequence representing steady oscillation in the system described by

$$\frac{dx}{dt} = y, \qquad \frac{dy}{dt} = \mu(1 - x^2)y - x^3 + B\cos\nu t \tag{5}$$

with $\mu = 0.2$, $B = 17$ and $\nu = 4$. These points are plotted after the transient state has decayed. Even if many more images are plotted than shown in the figure, the movement of images is not periodic. The set of points resembles

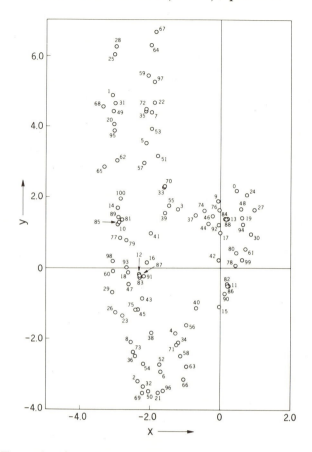

Fig. 6 Example of a point sequence representing steady oscillation which occurs in the system described by Eqs. (5), the parameters being $\mu = 0.2$, $B = 17$ and $\nu = 4$.

a bumpy ring pattern in the xy plane. The movement of images over long intervals of time is not reproducible, as with the previous examples. Figure 7 shows the global aspect of the fixed point, periodic points and invariant curves of the mapping T of this case. In this figure, the symbol $^{i}D_{j}^{n}$ indicates the i-th directly unstable n-periodic point, and the suffix j $(j = 1, 2, \cdots, n)$ represents the order of the successive movements of the images in the n-periodic group under the transformation T. Similar symbols are applied to the completely stable (S), completely unstable (U) and inversely unstable (I) periodic points and fixed points. As seen in this figure, there are four directly unstable 2-periodic points $^{1}D_{j}^{2}$ and $^{2}D_{j}^{2}$ and four inversely unstable 2-periodic points $^{1}I_{j}^{2}$ and $^{2}I_{j}^{2}$ $(j = 1, 2)$. The α-branches (heavy line) and ω-branches (thin line) of these points are distinguished by solid lines and dashed lines, respectively.

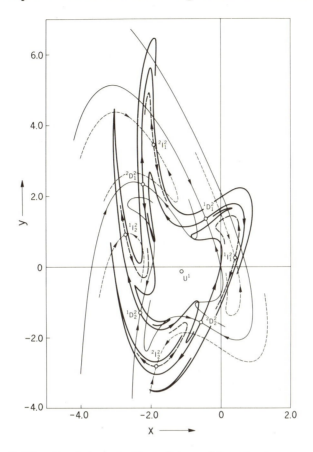

Fig. 7 Fixed point, periodic points and invariant curves of
the mapping for Eqs. (5), the parameters being
$\mu = 0.2$, $B = 17$ and $\nu = 4$.

Also, these α-branches are asymptotic to each other, forming a ring-like domain enclosing the completely unstable fixed point U^1. The image P_n representing the steady state in Fig. 6 continues to move in the neighborhood of these α-branches.

Finally, let us examine a more complicated example. Figure 8 shows a point sequence representing the steady oscillation of the images P_n in the system described by Eqs. (5) with $\mu = 0.2$, $B = 1.8$ and $\nu = 0.6$. Figure 9 shows the global phase-plane portrait for this case. As seen in these figures, there exist no completely stable or completely unstable points in this system.

The numerical examples given above were obtained by use of analog and digital computers. In the course of computer experiments, special attention was paid to the following points.

107

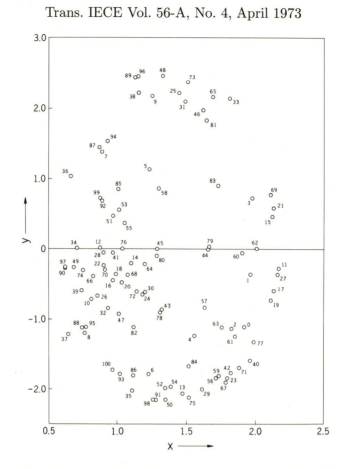

Fig. 8 Example of a point sequence representing steady oscillation
which occurs in the system described by Eqs. (5), the parameters
being $\mu = 0.2$, $B = 1.8$ and $\nu = 0.6$.

1. In the analog simulation, the amplitude of the external sinusoidal
   force must be kept constant.

2. In the digital simulation, considering that the Eq. (1) is a periodic
   system with the period L, the numerical integration has been carried
   by choosing the step size h to be an integral submultiple $h = L/N$
   of the period L for the interval $0 \leq t \leq Nh$, and by iterating this
   procedure.

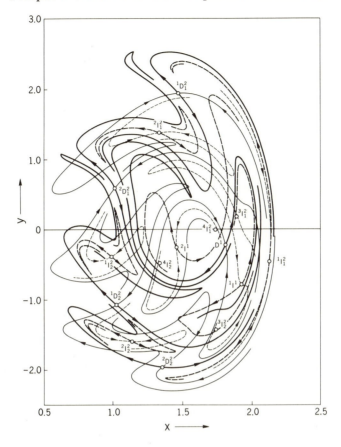

Fig. 9 Fixed points, periodic points and invariant curves of the mapping
for Eqs. (5), the parameters being $\mu = 0.2$, $B = 1.8$ and $\nu = 0.6$.

3. DISCUSSION

In the preceding section non-periodic steady oscillations have been observed
in computer simulated solutions of the forced oscillatory system (3) and of the
forced self-oscillatory system (5).[†] These differential equations are of class D,
i.e., dissipative systems for large displacements, and consequently have maximum finite invariant sets.[‡] This set is a positively asymptotically stable connected closed set and its configuration represents the behavior of the solutions of

---

[†]The systems treated in the preceding section are supposed to be structurally stable. Hence,
the systems considered in this section are assumed to possess this structural stability.

[‡]The area of the maximum finite invariant set is zero for the transformation T defined by
Eqs. (3) with $k > 0$.

the differential equation for $t \to \infty$.[§] However, all points of the maximum finite invariant set do not always represent steady solutions and the set may contain points representing transient states. Here, let us momentarily close our eyes to the stability required for the expression of realizable physical states, but let us define steady solutions in the mathematical sense as solutions corresponding to complete sequences on the xy plane which are stable according to Poisson; then all steady solutions can be expressed by the set of pseudo-recurrent points. The set of these pseudo-recurrent points belongs to the set of central points, which is obtained by subtracting the wandering points of all orders from the maximum finite invariant set. The resulting set includes all minimal sets representing recurrent motions, harmonic sets representing almost periodic motions and periodic sets representing periodic motions.

Here, let us consider the first example mentioned in Sec. 2.1. The maximum finite invariant set in Fig. 2 is the union of three fixed points S^1, D^1 and I^1 and the closures of α-branches of points D^1 and I^1. The set of non-wandering points within this set can be taken as the two fixed points S^1 and D^1 and the closure M of the α-branches of the point I^1.[¶] These are mutually separated invariant closed sets. Among them, the fixed point S^1 and the invariant closed set M are positively asymptotically stable. When a positively asymptotically stable invariant closed set such as the fixed point S^1 consists of a single periodic set, the periodic solution passing through the point S^1 is asymptotically stable according to Lyapunov. This stability permits the correction of errors and disturbances in the computer simulation to maintain the periodic oscillation in the system, and establishes that the periodic solution through S^1 is a physically realizable steady motion. Since the invariant closed set M is positively asymptotically stable, the image under the transformation T of any neighboring point of M is asymptotic to M and will not deviate from M. However, M contains numerous minimal sets, quasi-minimal sets and higher order wandering points, all of which appear to be unstable, and so it seems that any given solution passing through these points can not possess asymptotic stability individually. Hence, the phenomenon can only be explained by assuming that the totality of solutions contained in M represents the steady behavior of the computer-

---

[§]Let M be a bounded invariant closed set in the phase plane. A set M is called positively stable if, for any neighborhood U of the set M, there can be found a neighborhood V of the set M such that if $P \in V$, then $P_n (= T^n P,\ n > 0) \in U$. A set M is called positively asymptotically stable if M is positively stable and if there can be found a neighborhood V of the set M such that if $P \in V$, then $\Omega_P (= \omega\text{-limit set of } P) \subset M$. This stability is an extension of the concept of orbital (asymptotic) stability of the solutions of differential equations.

[¶]Whether all points in M are non-wandering points must be examined. Here, after consideration, M is assumed not to contain wandering points. If M contains wandering points, the set excluding them should be taken as M.

simulated system. That is, the representative point exhibiting the oscillatory state in the system continues to move randomly in the neighborhood of the numerous solutions (e.g. minimal sets) contained in M under the influence of errors, disturbances and minute variations of the system and so on. Therefore, although any particular solution within the invariant closed set M in the phase plane exhibits instability which is incompatible with the description of a physically realizable long-term steady behavior, still the set M as a whole represents the total configuration of the phenomenon in the phase plane. Considering this nature, let us call the oscillatory phenomenon exhibited by the solutions contained in the invariant closed set M a randomly transitional oscillation.

In the example given in Fig. 5, all points in the maximum finite invariant set are non-wandering and this set is a positively asymptotically stable invariant closed set representing the randomly transitional oscillation.

Next, let us consider the examples given in Sec. 2.2. The maximum finite invariant set in Fig. 7 is the connected closed region surrounded by the closure of the α-branches of the directly and inversely unstable 2-periodic points D^2 and I^2. In this case, the positively asymptotically stable invariant closed set consisting of non-wandering points is the invariant set obtained by excluding the completely unstable fixed point U^1 and its neighboring wandering points from the maximum finite invariant set. This set can be considered as the closure of the α-branches of the 2-periodic points D^2 and I^2. As seen in Figs. 6 and 7, the configuration of this set has the appearance of a bumpy ring domain enclosing the point U^1.

The last example given in Figs. 8 and 9 is so complicated that more detailed results than shown in the figures can not be expected from the simulation study.

As pointed out in this section, the steady oscillations in physical systems can be represented by positively asymptotically stable invariant closed sets of non-wandering points. However, when the action of this system which restores the displacements of the images deviated from the invariant closed set due to disturbances is weaker than the effect of these stochastic quantities, it becomes difficult to represent the actual phenomenon by this set. Such steady oscillations are frequently observed in the physical systems with small dissipation.

Theoretically, it is interesting to decompose the structure of a positively asymptotically stable invariant closed set of non-wandering points into the minimal sets and quasi-minimal sets contained in this invariant closed set, and to determine their numbers, properties and mutual relations, and also the asymptotic behavior of images under T in the neighborhood of these sets. However, simulation studies by their nature can give only partial and incomplete results in this direction.

Trans. IECE Vol. 56-A, No. 4, April 1973

4. CONCLUSION

In this paper, non-periodic steady oscillations have been observed in computer simulated systems described by second-order nonlinear ordinary differential equations. The equations under study are of class D, and should be structurally stable. It has been shown that, in spite of the deterministic character of the equations, there occur random oscillations in the computer simulated systems affected by stochastic quantities such as errors, external disturbances and so on. These non-periodic oscillations can be explained by the random movement of an actual representative point among solutions (e.g. minimal sets) contained in a positively asymptotically stable invariant closed set of non-wandering points. Each individual minimal set within the positively asymptotically stable invariant set is unstable, and no single minimal set can describe the steady oscillations which are observed. Further, it is proposed that this type of oscillation should be called randomly transitional oscillations.

It will be easily seen from the preceding discussion that, however accurate are the computers used, this type of random oscillation occurs inevitably so long as the computers are not ideal. This randomly transitional oscillation is typically observed in computer simulated systems of not only second-order nonlinear ordinary differential equations, but also multi-variable nonlinear equations. Further, the oscillations are supposed to be actual steady states in a wide range of nonlinear oscillatory systems such as electronic circuits containing nonlinear elements. Hence, from the point of view of the theory of nonlinear oscillations, even when apparently random oscillations are observed, deterministic nonlinear equations can still capture the essential nature of the phenomenon. Therefore, it can be considered that deterministic equations are appropriate mathematical models of such phenomenon.

REFERENCES

1. C. Hayashi, *Nonlinear Oscillations in Physical Systems.* McGraw-Hill, New York (1964).

2. C. Hayashi, Y. Ueda, N. Akamatsu and H. Itakura, On the behavior of self-oscillatory systems with external force, *Trans. Inst. Elec. Commun. Engrs* **53-A**, 150-158 (1970).

3. H. Poincaré, *Les Méthodes Nouvelles de la Mécanique Céleste.* Vol. 3, Chap. 33, Gauthier-Villars, Paris (1899).

4. G. D. Birkhoff, Surface transformations and their dynamical applications, *Acta Math.* **43**, 1-119 (1920); G. D. Birkhoff and P. A. Smith, Structure analysis of surface transformations, *J. Math.* (Liouville), S. 9, Vol. 7, pp. 345-379 (1928): in G. D. Birkhoff, *Collected Mathematical Papers.* Vol. 2, Dover Publications, New York (1968).

5. G. D. Birkhoff, *Dynamical Systems.* American Mathematical Society, Providence, RI (1927).

6. V. V. Nemytskii, Topological problems of the theory of dynamical systems, *Stability and Dynamic Systems.* American Mathematical Society, Translation series one, Vol. 5, pp. 414-497, Providence, RI (1927).

7. V. V. Nemytskii and V. V. Stepanov, *Qualitative Theory of Differential Equations.* Princeton University Press, Princeton, NJ (1960).

8. N. Levinson, Transformation theory of non-linear differential equations of the second order, *Ann. Math.* **45**, 723-737 (1944).

9. S. Furuya and J. Nagumo, *Theory of Nonlinear Oscillations* (*in Japanese*). Iwanami Kohza, Gendai Ohyo Sugaku B. 6-b, Iwanami Shoten, Tokyo (1957).

10. T. Saito, *Topological Dynamics* (*in Japanese*). Kyoritsu Kohza, Gendai no Sugaku 24, Kyoritsu Shuppan, Tokyo (1971).

APPENDICES

I. Review of Transformation Theory

The behavior of solutions of the differential equation (1) can be studied effectively by the topological methods originated by H. Poincaré and developed by G. D. Birkhoff and others. In this Appendix, the basic concepts of the transformation theory of differential equations are reviewed.

The first-order simultaneous equations rewritten from the differential equation (1) define a one-to-one, continuous and orientation-preserving transformation T of the phase plane into itself. The infinite sequence of points $\cdots$, P_{-2}, P_{-1}, P, P_1, P_2, $\cdots$ consisting of images P_n generated by applying the transformation T^n ($n = 0, \pm 1, \pm 2, \cdots$) on an arbitrary point P in the phase plane is called a complete sequence of P and a point sequence P, P_1, P_2, $\cdots$ is called a positive half-sequence of P. An accumulation point of the positive half-sequence of P is called an ω-limit point of P and the totality of these points is called an ω-limit set Ω_P of P. Similarly, for a negative half-sequence P, P_{-1}, P_{-2}, $\cdots$, α-limit point and α-limit set A_P of P can be defined. The union of a complete sequence and its α- and ω-limit sets is called a complete group.

When the image $TE = \{P;\ T^{-1}P \in E\}$ obtained by applying the transformation T to a point set E coincides with E itself, the set E is said to be invariant with respect to T. The above-mentioned α- and ω-limit sets and the complete group are examples of invariant closed sets.$^{\|}$

A point P is called positively stable according to Poisson if it is an ω-limit point of P, and negatively stable according to Poisson if it is an α-limit point of P. A point both positively and negatively stable according to Poisson is called stable according to Poisson. As seen in this definition, if P is positively stable according to Poisson, then every point of the complete sequence of P is also positively stable according to Poisson. A similar statement can be made for points negatively stable and stable according to Poisson.

Next, let us explain the concept of the set of central points introduced by G. D. Birkhoff. If an arbitrary connected region σ in the phase plane is not intersected by any of its images $\cdots, \sigma_{-2}, \sigma_{-1}, \sigma_1, \sigma_2, \cdots$ under the transformation T^n $(n = \pm 1, \pm 2, \cdots)$, σ is called a wandering region and its points wandering points. A point in the phase plane which is contained in no wandering region is called a non-wandering point. Among the non-wandering points are α- and ω-limit points. The totality of non-wandering points in the phase plane constitutes an invariant closed set M^1 with respect to T. Let us assume that M^1 is non-empty and is not identical with the phase plane itself** and let the complement of M^1 be denoted by W^1, then the points of W^1 are known to be asymptotic to M^1 on indefinite iteration of T or T^{-1}. Now let us take the set M^1 as fundamental instead of the entire plane. A connected region σ which contains points of M^1 is called wandering with respect to M^1 if the set $\sigma \cap M^1$ of points common to σ and M^1 is intersected by none of its images under the transformation T^n $(n = \pm 1, \pm 2, \cdots)$. The points of M^1 which are contained in such a region are called wandering with respect to M^1, and their totality is denoted by W^2. The set $M^2 = M^1 - W^2$ is a non-empty invariant closed set consisting of the points which are non-wandering with respect to M^1, and the points of W^2 tend toward M^2 asymptotically on indefinite iteration of T or T^{-1}. In case $M^1 = M^2$, M^1 is called non-wandering with respect to itself. In case M^2 is not identical with M^1, i.e. a proper subset of M^1, the process may be carried one step further yielding the set M^3 of points which are non-wandering with respect to M^2. Thus the process is continued reaching the set which is non-wandering with respect to itself. In case, however, that

$^{\|}$As for the maximum finite invariant sets, fixed points, periodic points, invariant closed curves, doubly asymptotic points, etc., refer to [2, 8]

**Since the transformation under consideration is defined by the differential equation of class D, the set M^1 of non-wandering points is included in the maximum finite invariant set and hence non-empty.

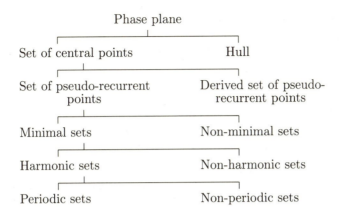

Fig. A Decomposition of a phase plane.

no such set appears after a finite number of steps, an infinite sequence M^1, M^2, $\cdots$ appears which satisfies $M^1 \supset M^2 \supset M^3 \supset \cdots$. In such a case, let us denote their intersection $\cap_{n=1}^{\infty} M^n$ by M^ω, then M^ω is a non-empty invariant closed set, to which the same process may be applied successively, yielding $M^{\omega+1}$, then $M^{\omega+2}, \cdots$. The sequence of sets thus obtained $M^1 \supset M^2 \supset \cdots \supset M^\omega \supset M^{\omega+1} \supset M^{\omega+2} \supset \cdots \supset M^{\omega 2} \supset M^{\omega 2+1} \supset \cdots$ is a well-ordered sequence and each set is non-empty and is a proper subset of all those preceding it. It is known that the sequence terminates with a definite ordinal r of Cantor's second ordinal class, i.e., $M^r = M^{r+1} = \cdots \neq \phi$ (ϕ: empty set). The points of M^r are called central points, and a complete sequence of central points is called a central motion. The set of points which is non-wandering with respect to itself, such as M^r, is said to possess the property of regional recurrence. Points of the sets W^2, W^3, $\cdots$ are called wandering points of higher orders, and a set of points outside the central points is called a hull.

In conclusion, let us decompose a set of central points. A point P is called pseudo-recurrent if it is both an α- and ω-limit point of its own complete sequence. As it has been proven that the set of central points is identical with the union of the set of pseudo-recurrent points and their derived set, a central point is either a pseudo-recurrent point or an accumulation point of pseudo-recurrent points. A pseudo-recurrent point is stable according to Poisson, and the complete group of a pseudo-recurrent point is called a quasi-minimal set. Specifically, a quasi-minimal set is called minimal if it is non-empty, closed and invariant, and has no proper subset possessing these three properties. A point of a minimal set is called recurrent, and the complete sequence of a recurrent point is called a recurrent motion. Further, if a complete sequence in a minimal set is almost periodic, then such a minimal set is called harmonic and points of a harmonic set are called almost periodic points. Finally, as the most limited case, if a complete sequence in a harmonic set is periodic, such a harmonic set is called periodic, and points of a periodic set are called periodic or fixed points. The above-mentioned items are summarized in Fig. A.

II. Supplements to the Numerical Examples

Details of the fixed and periodic points appearing in the numerical examples described in Secs. 2.1 and 2.2 are listed in Tables 1 and 2. The characteristic numbers in the Tables represent eigenvalues of the transformation matrix obtained by linearizing T or T^2 in the neighborhood of the fixed points. The numerical integration was performed on the FACOM 230-60 computer by using the RKG method with an integration step size of one-sixtieth of the period L. The authors wish to express their sincere thanks to the staff of the computer center at Kyoto University.

Table 1   Fixed points and periodic points of Eq. (3)
and their characteristic numbers

Parameter	Point	Fixed point, periodic point		Characteristic number	
		x_0	y_0	m_1	m_2
$k = 0.2$	S^1	0.639	1.085	$0.0994 \pm 0.524i$	
$B = 0.3$	D^1	−0.677	0.687	2.42	0.118
$B_0 = 0.08$	I^1	−0.154	0.103	−0.147	−1.94
$k = 0.2$	I^1	1.866	1.204	−0.198	−1.43
$B = 1.2$	I_1^2	1.230	1.979	−0.0291	−2.79
$B_0 = 0.85$	I_2^2	1.923	−0.380	−0.0291	−2.79

Table 2 Fixed points and periodic points of Eq. (5)
and their characteristic numbers

Parameter	Point	Fixed point, periodic point		Characteristic number	
		x_0	y_0	m_1	m_2
$\mu = 0.2$ $B = 17$ $\nu = 4$	U^1	−1.132	−0.142	−0.615 ± 0.861i	
	$^1D_1^2$	−0.421	1.345	7.35	0.0898
	$^1D_2^2$	−2.305	−1.294	7.35	0.0898
	$^2D_1^2$	−2.218	2.320	7.35	0.0898
	$^2D_2^2$	−0.578	−1.568	7.35	0.0898
	$^1I_1^2$	0.410	0.239	−0.108	−3.74
	$^1I_2^2$	−2.718	0.905	−0.108	−3.74
	$^2I_1^2$	−1.910	3.454	−0.108	−3.74
	$^2I_2^2$	−1.856	−2.809	−0.108	−3.74
$\mu = 0.2$ $B = 1.8$ $\nu = 0.6$	D^1	1.813	−0.214	22.3	0.0715
	$^1I^1$	1.934	−0.787	−0.775	−1.35
	$^2I^1$	1.465	−0.249	−0.775	−1.35
	$^1D_1^2$	1.478	1.943	10.4	0.0841
	$^1D_2^2$	1.028	−1.076	10.4	0.0841
	$^2D_1^2$	1.016	0.589	10.4	0.0841
	$^2D_2^2$	1.357	−1.962	10.4	0.0841
	$^1I_1^2$	2.136	−0.462	−0.285	−2.52
	$^1I_2^2$	0.999	−0.397	−0.285	−2.52
	$^2I_1^2$	1.347	1.372	−0.285	−2.52
	$^2I_2^2$	1.145	−1.603	−0.285	−2.52
	$^3I_1^2$	1.896	0.182	−0.0513	−29.4
	$^3I_2^2$	1.741	−1.442	−0.0513	−29.4
	$^4I_1^2$	1.741	−0.010	−0.0513	−29.4
	$^4I_2^2$	1.339	−0.481	−0.0513	−29.4

Selection 4

Int. J. Non-Linear Mechanics, Vol. 20, No. 5/6, pp. 481–491, 1985
Printed in Great Britain.

Pergamon Press Ltd.

RANDOM PHENOMENA RESULTING FROM NON-LINEARITY IN THE SYSTEM DESCRIBED BY DUFFING'S EQUATION†

Yoshisuke Ueda

Department of Electrical Engineering, Kyoto University, Kyoto 606, Japan

1. INTRODUCTION

In physical phenomena, uncertainties lie between causes and effects. When uncertain factors are small, their effects may be neglected in most physical systems and the phenomena under consideration are treated as deterministic ones. Whereas in non-linear systems on some conditions, however small uncertain factors may be, they sometimes cause global changes of state variables even in the steady states. These kind of phenomena originate from global structures of the solutions for non-linear equations describing physical systems. In such systems, steady motions may be observed exhibiting stochastic properties.

This paper deals with random oscillations which occur in a series-resonance circuit containing a saturable inductor under the impression of a sinusoidal voltage. Firstly, a differential equation is derived from the electrical circuit under discussion. The discrete dynamical system is introduced by using the solutions of the differential equation. The terminology used in the following descriptions is explained briefly. Secondly, experimental results concerning random oscillations are obtained by using analog and digital computers. Finally, the experimental results are examined and the problems arising from them are summarized.

The phenomenon treated in this paper should be called turbulence in electric circuits. A series of results obtained in this paper disclose an important feature of non-linear phenomena not only in electrical circuits but also in general physical systems.

2. STEADY OSCILLATIONS AND ATTRACTORS

2.1. *Differential equation*

A series-resonance circuit containing a saturable inductor is shown in Fig. 1. With the notation of the figure, the equation for the circuit is written as

$$\left. \begin{aligned} n\frac{d\phi}{dt} + Ri_R &= E\sin\omega t \\[2mm] Ri_R &= \frac{1}{C}\int i_c\,dt, \qquad i = i_R + i_c \end{aligned} \right\} \tag{1}$$

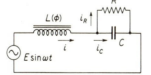

Fig. 1. Series-resonance circuit with non-linear inductance.

† This paper was translated by the author from his article in Japanese published in the *Transactions of the Institute of Electrical Engineers of Japan*, Vol. A98, March 1978, with kind permission of the Institute.

where n is the number of turns of the inductor coil, and ϕ is the magnetic flux in the core. Let us consider the case in which the saturation curve of the core is expressed by

$$i = a\phi^3 \tag{2}$$

It is to be noted that the effect of hysteresis is neglected in equation (2). Here we introduce the dimensionless variable x, defined by

$$\phi = \Phi_n x \tag{3}$$

where Φ_n is an appropriate base quantity of the flux and is fixed by the relation

$$n\omega^2 C\Phi_n = a\Phi_n^3 \tag{4}$$

Then, eliminating i_R and i_C in equation (1) and using equations (2), (3) and (4), we obtain the well-known Duffing's equation

$$\frac{d^2x}{d\tau^2} + k\frac{dx}{d\tau} + x^3 = B\cos\tau$$

where

$$\tau = \omega t - \tan^{-1}k, \qquad k = \frac{1}{\omega CR}$$

$$B = \frac{E}{n\omega\Phi_n}\sqrt{1 + k^2}. \tag{5}$$

2.2. Discrete dynamical system

Equation (5) is rewritten in simultaneous form as

$$\frac{dx}{d\tau} = y, \qquad \frac{dy}{d\tau} = -ky - x^3 + B\cos\tau \tag{6}$$

A discrete dynamical system on the xy plane is introduced by using the solutions of equation (6). In order to see this, first consider the solution $(x(\tau, x_0, y_0), y(\tau, x_0, y_0))$ of equation (6), which, when $\tau = 0$, is at the point $p_0 = (x_0, y_0)$ of the xy plane. Let $p_1 = (x_1, y_1)$ denote the point specified by $x_1 = x(2\pi, x_0, y_0), y_1 = y(2\pi, x_0, y_0)$; then either a C^∞-diffeomorphism f_λ

$$f_\lambda : R^2 \to R^2$$

$$p_0 \mapsto p_1$$

where

$$\lambda = (k, B) \in \Lambda \tag{7}$$

of the xy plane into itself or a discrete dynamical system on R^2 is defined. For the theory of dynamical systems, see refs. [1–6].

For the circuit with dissipation ($k > 0$), f_λ is of class D or a dissipative system for large displacements and is a contractive mapping of the xy plane into itself. This implies that an orbit $\text{Orb}(p) = \{f_\lambda^n(p)\,|\,n \in Z\}$ of the discrete dynamical system (7) which starts from an arbitrary point $p \in R^2$ is positively stable in the sense of Lagrange and that there exists positively asymptotically stable, f_λ-invariant, maximum compact set $\Delta(f_\lambda)$ with zero area. Therefore, investigation of the steady oscillations leads to examining the maximum compact set $\Delta(f_\lambda)$ on R^2 and the behavior of its neighboring orbits.

2.3. *Steady oscillations and attractors*

Since there exist uncertain factors, such as noise, in actual electric circuits, changes of voltages and/or currents are represented approximately by the solutions of the differential equations of the circuit. Accordingly, when the representative point of the circuit moves along an asymptotically stable solution, effects of noise may be neglected and deterministic phenomenon occurs. But stochastic phenomenon is caused when the representative point wanders, under the influence of noise, in the neighborhood of infinite solutions.

In the following, let us define the attractor as asymptotically stable, f_λ-invariant, compact set on the xy plane which exhibits a steady oscillation sustained in the actual circuit of Fig. 1. An attractor exhibiting a periodic oscillation is either a fixed point or a periodic group of the discrete dynamical system. An attractor exhibiting a random oscillation is considered to be an f_λ-invariant compact set containing infinite minimal sets.

3. EXPERIMENTS ON THE RANDOM OSCILLATIONS

In this section, experimental results obtained by using analog and digital computers are shown. Simulation and/or calculation errors are unavoidable in the computer solutions for the differential equation. Therefore, random quantities are not introduced intentionally but these errors are regarded as uncertainties acting on the system. These errors seem to be sufficiently small compared with noises in the actual circuit.

In the electric circuit as shown in Fig. 1, random oscillations can be observed within some intervals of the applied voltage and the circuit's constants. In fact, they occur in the ranges $B = 9.9$–13.3 for $k = 0.1$ and $k = 0$–0.31 for $B = 12.0$. In the following, experimental results are given for the steady oscillations, which occur in the computer-simulated systems for the representative values of the system parameters

$$\lambda = (k, B) = (0.1, 12.0) \tag{8}$$

3.1. *Waveforms*

Figure 2 shows two waveforms of the steady states sustained in the analog computer-simulated system. Figure 2(a) shows a periodic oscillation containing remarkable higher harmonic components. The upper waveform is the applied voltage $B \cos \tau$ and the lower one is the normalized magnetic flux $x(\tau)$. Figure 2(b) shows a random oscillation. This waveform is not reproducible in every analog computer experiment. In the digital simulation, different waveforms are observed depending on the integration method and the step size. Therefore, the waveform of Fig. 2(b) is a realization of the random process $\{X(\tau)\}$.

3.2. *Phase-plane analysis*

Figure 3 shows a long-term orbit (a realization) of a random oscillation. The movement of images under iterations of f_λ is not uniquely determined even for the same initial point, but

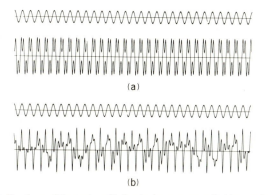

(a)

(b)

Fig. 2. Waveforms of the steady oscillations in the system prescribed by equation (8).

123

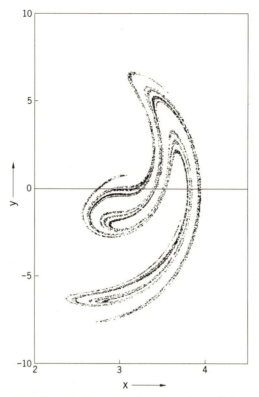

Fig. 3. Observed orbit corresponding to the random oscillation.

the general aspect (location, shape and size) of the orbit is reproducible, and further it seems stable in the Poisson sense. Therefore, a set of points as shown in Fig. 3 should be regarded as an outline of an attractor M representing the random oscillation.

Figure 4 shows fixed points $^1D^1$ (directly unstable), $^1I^1$ and $^2I^1$ (inversely unstable), and outlines of unstable $W^u(^1D^1)$ and stable $W^s(^1D^1)$ manifolds of $^1D^1$. If the unstable manifolds (thick lines in the figure) are prolonged, they tend to the orbit of Fig. 3. In other words, the closure of unstable manifolds of $^1D^1$, i.e. $ClW^u(^1D^1)$, is regarded as the attractor M for the random oscillation.

Figure 5 shows the global phase-plane structure of the diffeomorphism f_λ. The stable manifolds (thick lines) of the saddle $^2D^1$ (directly unstable fixed point) are the boundary of the two domains of attraction, and the sink S^1 is the attractor corresponding to the periodic oscillation of Fig. 2(a).

3.3. Spectral analysis

In this section, spectral analysis of the random oscillation is shown. To this end, we regard the random process $\{X(\tau)\}$ as the periodic random process $\{X_T(\tau)\}$ with a sufficiently long period T, where T is a multiple of 2π. That is, let $x_T(\tau)$ be a periodic function with period T, which coincides with a realization $x(\tau)$ of $\{X(\tau)\}$ in the interval $(-T/2, T/2]$. Then a realization $x_T(\tau)$ is expanded into Fourier series as

$$x_T(\tau) = \frac{a_0}{2} + \sum_{m=1}^{\infty} (a_m \cos m\omega_0\tau + b_m \sin m\omega_0\tau), \qquad \omega_0 = \frac{2\pi}{T} \tag{9}$$

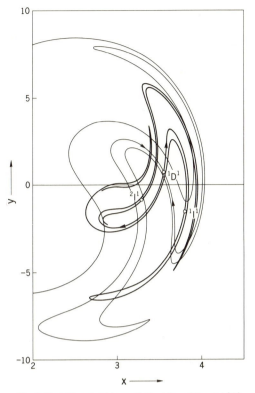

Fig. 4. Unstable and stable manifolds of the saddle point $^1D^1$.

where

$$
\left.
\begin{aligned}
a_m &= \frac{2}{T} \int_{-T/2}^{T/2} x_T(\tau) \cos m\,\omega_0 \tau \, \mathrm{d}\tau \\[2mm]
b_m &= \frac{2}{T} \int_{-T/2}^{T/2} x_T(\tau) \sin m\,\omega_0 \tau \, \mathrm{d}\tau \\[2mm]
m &= 0, 1, 2, \dots
\end{aligned}
\right\}
\tag{10}
$$

Fourier's coefficients a_m and b_m are random variables because $x_T(\tau)$ is a realization of the random process $\{X_T(\tau)\}$. From these coefficients, the mean value $m_X(\tau)$ and the average power spectrum $\Phi_X(\omega)$ of the random process $\{X(\tau)\}$ can be estimated as follows.

$$
\begin{aligned}
m_X(\tau) &= \langle X(\tau) \rangle = \lim_{T \to \infty} \langle X_T(\tau) \rangle \\[2mm]
&\doteq \langle X_T(\tau) \rangle = \left\langle \frac{a_0}{2} \right\rangle + \sum_{m=1}^{\infty} \left[\langle a_m \rangle \cos m\,\omega_0 \tau + \langle b_m \rangle \sin m\,\omega_0 \tau \right]
\end{aligned}
\tag{11}
$$

$$
\left.
\begin{aligned}
\Phi_X(\omega) &= \lim_{T \to \infty} \left\langle \frac{1}{T} \left| \int_{-T/2}^{T/2} x_T(\tau) \mathrm{e}^{-i\omega\tau} \, \mathrm{d}\tau \right|^2 \right\rangle \\[2mm]
&\doteq \Phi_X(m\,\omega_0) = \frac{2\pi}{\omega_0} \left\langle \frac{1}{4}(a_m^2 + b_m^2) \right\rangle \\[2mm]
\omega_0 &= \frac{2\pi}{T}
\end{aligned}
\right\}
\tag{12}
$$

125

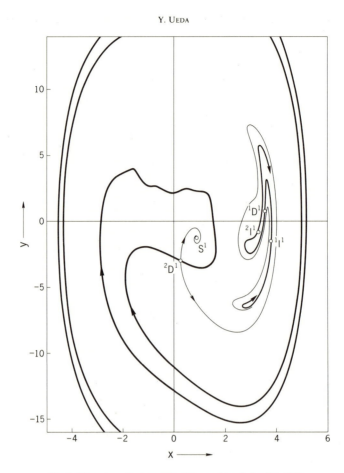

Fig. 5. Phase-plane structure of the diffeomorphism f_λ, $\lambda = (0.1, 12.0)$.

The ensemble average is calculated by regarding successive waveforms at the intervals $((n - 1/2)T, (n + 1/2)T]$ $(n = 0, 1, 2, ..., N_s)$ as sample processes of $\{X_T(\tau)\}$.

Let us give the results thus estimated for the specified values of the system parameters given by equation (8). The mean value of $\{X(\tau)\}$ is given by

$$
\begin{aligned}
m_X(\tau) = \ & 1.72 \cos \tau + 0.22 \sin \tau \\
& + 1.21 \cos 3\tau - 0.26 \sin 3\tau \\
& + 0.25 \cos 5\tau - 0.06 \sin 5\tau \\
& + 0.07 \cos 7\tau - 0.02 \sin 7\tau \\
& + 0.02 \cos 9\tau - 0.01 \sin 9\tau
\end{aligned}
\tag{13}
$$

The mean value is found to be a periodic function. This indicates that the random process $\{X(\tau)\}$ is a non-stationary random process. Figure 6 shows the average power spectrum estimated by using equation (12). In the figure, line spectra at $\omega = 1, 3, 5, ...$ indicate the periodic components of the mean value as given by equation (13), and numerical values attached to line spectra represent the power concentrated on those frequencies. In every

126

computer experiment, the general aspect (location, shape and size) of the average power spectrum is reproducible. The average power of the random process $\{X(\tau)\}$ is given by

$$\lim_{T \to \infty} \frac{1}{T} \int_{-T/2}^{T/2} \langle X_T^2(\tau) \rangle \, d\tau = \frac{1}{T} \int_{-T/2}^{T/2} \langle X_T^2(\tau) \rangle \, d\tau = 3.08 \qquad (14)$$

3.4. *Spectral decomposition of the power*

It is easily seen that, due to the non-linearity of the inductor, the results of Fig. 6 and equation (14) in the preceding section do not have dimension of the electric power. Therefore, let us here examine the situation of spectral dispersion of electric power supplied by the source of single frequency. The terminal voltage of the capacitor

$$v(\tau) = \frac{B}{\sqrt{1 + k^2}} \sin(\tau + \tan^{-1} k) - y(\tau) \qquad (15)$$

is also a random process $\{V(\tau)\}$, and the mean value is given by

$$\begin{aligned}
m_V(\tau) = \; & 0.97 \cos \tau + 13.60 \sin \tau \\
& + 0.77 \cos 3\tau + 3.63 \sin 3\tau \\
& + 0.28 \cos 5\tau + 1.27 \sin 5\tau \\
& + 0.16 \cos 7\tau + 0.51 \sin 7\tau \\
& + 0.06 \cos 9\tau + 0.15 \sin 9\tau \\
& + 0.02 \cos 11\tau + 0.04 \sin 11\tau \\
& + 0.00 \cos 13\tau + 0.01 \sin 13\tau
\end{aligned} \qquad (16)$$

Figure 7 shows an average power spectrum of $\{V(\tau)\}$.† The average power of the random process is given by

$$\lim_{T \to \infty} \frac{1}{T} \int_{-T/2}^{T/2} \langle V_T^2(\tau) \rangle \, d\tau \doteqdot \frac{1}{T} \int_{-T/2}^{T/2} \langle V_T^2(\tau) \rangle \, d\tau = 104 \qquad (17)$$

From these results, spectral decomposition of the electric power dissipated in the shunt resistance with the capacitor $kv^2(\tau)$ is calculated. That is, the power supplied by the single frequency is decomposed and dissipated in the resistor at the following rate of frequencies:

fundamental component	89.3%
third harmonic component	6.6%
random component	3.1%
fifth harmonic component	0.8%

The results of this section are obtained by using analog and digital computers. In the analog computer experiments, various types of non-linear elements and a number of operating speeds (time scales) are used. In the digital computer experiments, integration

---

† In the present discussion, voltage, current and time are normalized by $n\omega\Phi_n$, $a\Phi_n^3$ and $1/\omega$, respectively. In this case, the source voltage is given by

$$\frac{B}{\sqrt{1 + k^2}} \sin(\tau + tan^{-1} k) = 1.19 \cos \tau + 11.88 \, sin \, \tau$$

and the mean value of the current by

$$\langle X^3(\tau) \rangle = 13.70 \cos \tau + 0.39 \sin \tau + \cdots$$

Therefore, the power supplied by the source turns out to be 10.4.

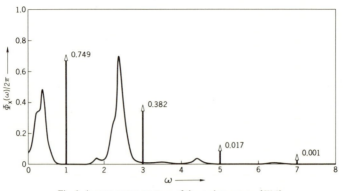

Fig. 6. Average power spectrum of the random process $\{X(\tau)\}$.

methods of Runge–Kutta–Gill and of Hamming with various step sizes are used, and further experiments have also been executed for both single and double precisions. It is confirmed that these results agree well with each other not only qualitatively but also quantitatively.

Fourier's transformation, equation (10), has been carried out by applying the discrete FFT algorithm

$$
\left.
\begin{aligned}
a_m &= \frac{1}{N} \sum_{k=-(N-1)}^{N} x\left(k\frac{T}{2N}\right) \cos\left(mk\frac{\pi}{N}\right), \\
&\qquad m = 0, 1, 2, \ldots, N \\
b_m &= \frac{1}{N} \sum_{k=-(N-1)}^{N} x\left(k\frac{T}{2N}\right) \sin\left(mk\frac{\pi}{N}\right), \\
&\qquad m = 1, 2, \ldots, N-1
\end{aligned}
\right\}
\tag{18}
$$

for $2N$ sampled values $x(\tau_k)$ at $\tau_k = kT/2N$ ($k = -(N-1), \ldots, N$). In the calculations, taking the errors related to the approximations of continuous variables by discrete variables and of infinite interval by finite interval into account, the values $T = 2\pi \times 2^{10}$, $2N = 2^{15}$ are used, and the ensemble average is estimated for 100 ($N_s = 99$) realizations.

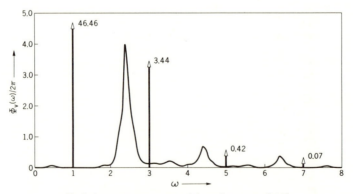

Fig. 7. Average power spectrum of the random process $\{V(\tau)\}$.

4. DISCUSSION OF THE EXPERIMENTAL RESULTS

In the present section, the experimental results given in the preceding section are discussed and the problems arising from them are summarized.

(1) The random oscillation is not a special one which appears only for particular values of the system parameters, but can be observed in a rather wide range of values. From Figs. 3 and 4 the attractor M of the random oscillation specified by the parameter values of equation (8) is regarded as a closure of unstable manifolds of the directory unstable fixed point $^1D^1$ of f_λ, i.e. $M = ClW^u(^1D^1)$. An appearance of the attractor seems to change continuously when the parameters are varied in the neighborhood of $\lambda = (0.1, 12.0)$. The movement of images in the attractor M under iterations of f_λ is not reproducible, but seems to be stable in the Poisson sense.

(2) The stable manifolds $W^s(^1D^1)$ intersect the unstable manifolds $W^u(^1D^1)$ forming a homoclinic cycle. As we see in Fig. 4, most intersections (doubly asymptotic points) are transversal, but, as is seen in the neighborhood of the point $(2.8, -2.0)$ of Fig. 4, it is expected that there exist some tangential points (doubly asymptotic points of the special type) on the prolonged manifolds. This fact suggests that the structure of the attractor may be unstable in the sense of Andronov–Pontryagin.

Existence of homoclinic points indicates that the attractor M contains infinite periodic groups. As shown by experiments, every periodic group is unstable. This fact implies that, even if sinks exist, their domains of attraction are so narrow that they are subject to the perturbations by the uncertainties acting on the system.

(3) The observed orbit $\{X(2n\pi), Y(2n\pi)\}$ $(n \in Z^+)$ of the discrete dynamical system, in other words, stroboscopic sequence of the computer solution with the same period as that of the periodic forcing seems to be a 2-dimensional stationary sequence taking values in the attractor.

(4) The random process $\{X(\tau)\}$ can be regarded as a sample process of the periodic non-stationary random process. Therefore, the mean value $m_X(\tau)$ is a periodic function with period 2π and the correlation function of the random component $\{R(\tau)\} = \{X(\tau) - m_X(\tau)\}$ is invariant under the periodic translations: $\tau \to \tau + 2n\pi$ $(n \in Z)$ [7]–[9].

In the computer experiments, an outline of the average power spectrum of the random process is reproducible. This indicates that the average power spectrum of the process is characterized by the structure of solutions regardless of the nature of uncertain factors. In other words, the simulation and/or calculation errors are not amplified into a random process but only bring about randomness in the phenomenon.

(5) From the above facts, the genesis and the properties of the random oscillation are summarized as follows. "The representative point of the actual system (which is not prescribed by the solution of the differential equation in the mathematical sense) continues to transit randomly among the infinitely many solutions due to the perturbations by uncertain factors of the system. The average power spectrum of the random oscillation depends practically not on the nature of uncertain factors but on the structure of the solutions emanating from the attractor".

In succession, the attractor representing random oscillations should be defined appropriately by "the asymptotically stable, compact, f_λ-invariant set which contains infinitely minimal sets connected to one another by the influence of uncertain factors in the actual system".

Because of those aforementioned, we have called this type of oscillation "the randomly transitional oscillation" [10]. The difference between randomly transitional oscillations and almost periodic oscillations is explained as follows. The attractor of the former is composed of infinite minimal sets, whereas that of the latter is made up of a single minimal set.

The problems arising from the above matters are summarized as follows.

(6) Let $\Omega(f_\lambda)$ be a set of non-wandering points in the domain of attraction for the attractor $M = ClW^u(^1D^1)$. Is $\Omega(f_\lambda)$ identical with M, or a proper subset of M? Does $\Omega(f_\lambda)$ contain minimal sets different from periodic groups? Does $\Omega(f_\lambda)$ contain minimal sets different from periodic groups? How is $\Omega(f_\lambda)$ decomposed?

(7) From the above item (1), the attractor M seems to be structurally stable in some sense. What is the concept of structural stability? That is, what kinds of space (of differential equations) and topology are used for the discussion of structural stability?

(8) How does the transition probability of the stroboscopic sequence $\{X(2n\pi),\ Y(2n\pi)\}$ $(n \in Z^{+})$ and the stochastic properties of $\{X(\tau)\}$ depend on the nature of uncertain factors of the system?

As mentioned in item (4), an average power spectrum scarcely depends on the simulation and/or calculation errors but is determined from the structure of the solutions emanating from the attractor. This fact seems applicable to general electric circuits provided that the uncertain factors are sufficiently small and have no special characteristics. For the case in which this conjecture does not hold, namely, for the case in which some kind of resonance may be expected, what kind of relationship is expected between the nature of the noise and the structure of the solutions passing the attractor? Under the influence of random noise having the characteristics above, the phenomenon must be discussed by introducing random parameters into the differential equations describing the electric circuit.

5. CONCLUSION

In the present paper, random phenomena resulting from non-linearity have been studied in the series-resonance circuit containing a saturable inductor. As a result of this investigation, a part of the genesis and of the stochastic properties of the random oscillation has been first clarified. This phenomenon should be called turbulence in electric circuits.

Although an example of randomly transitional phenomena has been studied in detail for the system described by Duffing's equation, this kind of phenomena have been observed in another non-linear system. Hence they may be regarded as general steady phenomena in non-linear systems [10]. Further, it seems interesting to examine the phenomena in reference to the turbulence in fluid dynamics [11].

The unsolved problems (6)–(8) pointed out in the preceding section relate closely to the global structure of solutions of differential equations both in time and in space. They also relate to uncertain factors of actual systems. They are really fundamental and difficult problems. It is hoped that these problems will deserve attention as material for further study.

Acknowledgements—The author wishes to express his sincere thanks to Professor Michiyoshi Kuwahara and Professor Chikasa Uenosono, both of Kyoto University for their thoughtful consideration and encouragement. He is likewise grateful to Professor Ken-ich Shiraiwa of Nagoya University and Professor Hisanao Ogura of the Kyoto Institute of Technology for their many useful comments and generous advice. The author also appreciates the assistance he received from Associate Professors Hiroshi Kawakami and Norio Akamatsu, both of Tokushima University, Miss Keiko Tamaki and Miss Yuriko Yamamoto, both of Kyoto University.

(The manuscript was received 30 June 1977, and the revised one 30 September 1977 by the Inst. Elect. Engrs. of Japan.)

REFERENCES

1. K. Shiraiwa, *Theory of Dynamical Systems.* Iwanami Shoten (1974).
2. N. Levinson, *Annls Math.* **45**, 723 (1944); **49**, 738 (1948).
3. V. V. Nemytskii and V. V. Stepanov, *Qualitative Theory of Differential Equations.* Princeton University Press, Princeton (1960).
4. G. D. Birkhoff, *Collected Mathematical Papers.* Dover, New York (1968).
5. S. Smale, *Bull. Am. Math. Soc.* **73**, 747 (1967).
6. Z. Nitecki, *Differentiable Dynamics.* MIT Press, Cambridge, Mass. (1971).
7. R. L. Stratonovich, *Topics in the Theory of Random Noise*, Gordon and Breach, New York (1963).
8. H. Ogura, *Trans. Inst. Elect. Commun Engrs. Japan.* **53-C**, 133 (March 1970).
9. H. Ogura, *IEEE Trans. Inf. Theory*, **IT-17**, 143 (1971).
10. Y. Ueda *et al., Trans. Inst. Elect. Commun Engrs, Japan.* **56-A**, 218 (April 1973).
11. J. E. Marsden and M. McCracken, *The Hopf Bifurcation and Its Applications.* Springer, Berlin (1976).

POSTSCRIPT

I deem it a great honour to be given the opportunity to translate my article into English and I would like to express my thanks to the members of the editorial board. In the following I am writing down some comments and fond memories of days past when I was preparing the manuscript with tremendous difficulty.

It was on 27 November 1961 when I met with chaotic motions in an analog computer simulating a forced self-oscillatory system. Since then my interest has been held by the phenomenon, and I have been fascinated by the problem "What are steady states in non-linear systems?". After nearly ten years, I understood "randomly transitional phenomenon", I published my findings in the *Transactions of the Institute of Electronics and Communication Engineers of Japan*, Vol. 56, April 1973 [10]. My paper then received a number of unfavorable criticisms from some of my colleagues: such as, "Your results are of no importance because you have not examined the effects of simulation and/or calculation errors at all", "Your paper is of little importance because it is merely an experimental result", "Your result is no more than an almost periodic oscillation. Don't form a selfish concept of steady states", and so forth. Professor Hiromu Momota of the Institute of Plasma Physics was the first to appreciate the worth of my work. He said "Your results give an important feature relating to stochastic phenomena" on

3 March 1974. Through his good offices I joined the Collaborating Research Program at the Institute of Plasma Physics of Nagoya University. These events gave me such unforgettable impressions that I continued the research with tenacity. At this moment I yearn for those days with great appreciation for their criticisms and encouragements.

By the middle of the 1970s, I had obtained many data of strange attractors for some systems of differential equations, but I had no idea to what journals and/or conferences I might submit these results. I was then lucky enough to meet with Professor David Ruelle who was visiting Japan in the early summer of 1978. He advised me to submit my results to the *Journal of Statistical Physics* [P1]. Further, he named the strange attractor of Fig. 3 "Japanese Attractor" and introduced it to the whole world [P2–P5]. At that time chaotic behavior in deterministic systems began to come under the spotlight in various fields of natural sciences. I fortunately had several opportunities to present my accumulated results [P6–P11]. It is worth while mentioning that, due to the efforts of Professor David Ruelle and Professor Jean-Michel Kantor, the Japanese Attractor will be displayed at the National Museum of Sciences, Techniques and Industries which will open in Paris, 1986. In these circumstances this paper is a commemorative for me and I sincerely appreciate their kindness on these matters.

As the reader will notice in this translation and also in ref. [P1] I was rather nervous of using the term "strange attractor", because I had no understanding of its mathematical definition in those days. Although I do not think I fully understand the definition of it even today, I begin to use the term "strange attractor" without hesitation because it seems to agree with reality. However, it seems to me that the term "chaos", though it is short and simple, is a little bit exaggerated. In the universe one does have a lot more complicated, mysterious and incomprehensible phenomena! I should be interested in readers' views of my opinion.

REFERENCES TO POSTSCRIPT

P1. Y. Ueda, *J. statist. Phys.* **20**, 181 (1979).
P2. D. Ruelle, *La Recherche*, **11**, 132 (1980).
P3. D. Ruelle, *The Mathematical Intelligencer* **2**, 126 (1980).
P4. D. Ruelle, *Mathematics Calendar*. Springer, Berlin (November 1981).
P5. D. Ruelle, *Czech. J. Phys.* **A32**, 99 (1982).
P6. Y. Ueda, New approaches to non-linear problems in dynamics, *SIAM J. appl. Math.* 311 (1980).
P7. Y. Ueda, *Annls N.Y. Acad. Sci.* **357**, 422 (1980).
P8. Y. Ueda and N. Akamatsu, *IEEE Trans Circuits and Systems*, **28**, 217 (1981).
P9. H. Ogura *et al.*, *Prog. theoret. Phys.* **66**, 2280 (1981).
P10. Y. Ueda, *Proc. 24th Midwest Symposium on Circuits and Systems*, p. 549. University of New Mexico (1981).
P11. Y. Ueda and H. Ohta, *Chaos and Statistical Methods*, p. 161. Springer, Berlin (1984).

Selection 5

EXPLOSION OF STRANGE ATTRACTORS EXHIBITED
BY DUFFING'S EQUATION

Yoshisuke Ueda

Department of Electrical Engineering
Kyoto University
Kyoto 606, Japan

INTRODUCTION

Various fascinating phenomena occur in nonlinear systems.[1] They are discussed generally by making use of steady solutions of a differential equation. The equation is usually derived from a real system by neglecting small uncertain factors—hence, it is deterministic. If a single solution of the equation is asymptotically stable and its basin is large compared with random noise, the corresponding phenomenon turns out to be deterministic. But if a bundle of solutions containing infinitely many unstable periodic solutions is asymptotically orbitally stable, a chaotic phenomenon appears, which results from the small uncertain factors in the real system. That is, the representative point of the physical state wanders chaotically in the bundle of solutions. Because of this characteristic, we have called the phenomenon a "chaotically transitional process."[2-5]*

We have long been studying this subject in connection with forced oscillatory phenomena in nonlinear electrical and electronic circuits. As analytical solutions of differential equations cannot be expected in these problems, we have been relying on analog and digital computers. Thus, both the global structure of solutions and the long-term movement of a representative point have been examined. So the question arises, Are computer solutions valid? However, we have not entered into details of this problem, though we have found consistency between the analog and digital results. Therefore, our results may lack mathematical rigor; nevertheless, they will have an important influence on many researchers in various fields.

This paper also describes the chaotically transitional processes exhibited by Duffing's equation. Special attention is directed toward the transition of the processes and the explosion of strange attractors is clarified.

PRELIMINARIES

This report deals with Duffing's equation,

$$\frac{d^2x}{dt^2} + k\frac{dx}{dt} + x^3 = B_0 + B_1 \cos t \tag{1}$$

---

*At first, we used the term "random" instead of "chaotic." This revision is due to the advice of Professor Joseph Ford of the Georgia Institute of Technology.

0077-8923/80/0357-0422 $01.75/1 © 1980, NYAS

or

$$\frac{dx}{dt} = y, \qquad \frac{dy}{dt} = -ky - x^3 + B_0 + B_1 \cos t. \qquad (2)$$

In this and the following sections, we present a brief review concerned with the chaotically transitional processes.

Discrete Dynamical System

A discrete dynamical system on the xy plane is introduced by using the solutions of equation 2. In order to see this, first consider the solution $(x(t, x_0, y_0), y(t, x_0, y_0))$ of (2), which, when $t = 0$, is at the point $p_0 = (x_0, y_0)$ of the xy plane. Let $p_1 = (x_1, y_1)$ denote the point specified by $x_1 = x(2\pi, x_0, y_0)$, $y_1 = y(2\pi, x_0, y_0)$; then either a C^∞-diffeomorphism f_λ,

$$f_\lambda : R^2 \to R^2$$
$$p_0 \mapsto p_1, \qquad (3)$$
$$\lambda = (k, B_0, B_1),$$

of the xy plane into itself or a discrete dynamical system on R^2 is defined.

Bundle of Solutions and Strange Attractors

Let us consider the case in which chaotic motion takes place in the real system exhibited by (2). This chaotically transitional process, $\{X(T)\}$, is represented by a bundle of solutions in the txy space that is asymptotically orbitally stable and contains infinitely many unstable periodic solutions. The set of points on the xy plane consisting of the cross section of the bundle at $t = 2n\pi$ $(n \in Z)$ is called a strange attractor.

Average Power Spectrum

By assuming that the process $\{X(t)\}$ is the periodic random process $\{X_T(t)\}$ with a sufficiently long period T, where T is a multiple of 2π, we can estimate the mean value $m_X(t)$ and the average power spectrum $\Phi_X(\omega)$ of $\{X(t)\}$ as follows:

$$m_X(t) = \langle X(t) \rangle \simeq \langle X_T(t) \rangle$$

$$= \left\langle \frac{a_0}{2} \right\rangle + \sum_{m=1}^{\infty} \{\langle a_m \rangle \cos m\omega_0 t + \langle b_m \rangle \sin m\omega_0 t\}, \qquad \omega_0 = \frac{2\pi}{T}, \quad (4)$$

$$\Phi_X(\omega) = \lim_{T \to \infty} \left\langle \frac{1}{T} \left| \int_{-T/2}^{T/2} x_T(t) e^{-i\omega t} \, dt \right|^2 \right\rangle \simeq \Phi_X(m\omega_0)$$

$$= \frac{2\pi}{\omega_0} \left\langle \frac{1}{4} (a_m^2 + b_m^2) \right\rangle, \qquad \omega_0 = \frac{2\pi}{T}, \qquad m = 0, 1, 2, \ldots . \quad (5)$$

where a_m and b_m are Fourier coefficients of a realization $x_T(t)$ of the process $\{X_T(t)\}$, i.e.,

$$x_T(t) = \frac{1}{2} a_0 + \sum_{m-1}^{\infty} \{a_m \cos m\omega_0 t + b_m \sin m\omega_0 t\}, \qquad \omega_0 = \frac{2\pi}{T},$$

$$a_m = \frac{2}{T} \int_{-T/2}^{T/2} x_T(t) \cos m\omega_0 t \, dt, \qquad m = 0, 1, 2, \ldots , \quad (6)$$

$$b_m = \frac{2}{T} \int_{-T/2}^{T/2} x_T(t) \sin m\omega_0 t \, dt, \qquad m = 1, 2, \ldots .$$

Exponentlike Quantities

One of the stochastic properties of the strange attractors is estimated by the exponentlike quantities e_u and e_s. To make this clear, let us calculate the characteristic roots m_{1i}, m_{2i} of $Df_\lambda(p_i)$, the derivative of f_λ evaluated at $p = p_i$, where $p_i = f_\lambda^i(p_0)$, $(i \in Z^+)$. Then, by setting

$$e_{un} = \frac{1}{2n\pi} \sum_{i-0}^{n-1} \ln \rho_{ui}, \qquad e_{sn} = \frac{1}{2n\pi} \sum_{i-0}^{n-1} \ln \rho_{si}, \quad (7)$$

where $\rho_{ui} = \max \{|m_{1i}|, |m_{2i}|\}$, $\rho_{si} = \min \{|m_{1i}|, |m_{2i}|\}$, we can define the exponentlike quantities by taking the limits

$$e_u = \lim_{n \to \infty} e_{un}, \qquad e_s = \lim_{n \to \infty} e_{sn}. \quad (8)$$

From our experiments, we see that the limits seem to exist and to be independent of p_0. According to the above definition, one of them, e_u, indicates the rate of divergence of nearby points in the attractor and the other, e_s, indicates the rate of attraction of the strange attractor under f_λ.

Proceeding in the same manner as above, exponentlike quantities e_u^j, e_s^j under every jth iteration of the mapping f_λ can be defined. These quantities, e_u^j and e_s^j, indicate the rates of divergence and attraction under the mapping f_λ^j, respectively. The following relation can be easily proved.

$$e_u^j + e_s^j = -k. \quad (9)$$

MAIN RESULTS AND REMAINING UNSOLVED PROBLEMS RELATING
TO THE CHAOTICALLY TRANSITIONAL PROCESSES
OF THE PREVIOUS INVESTIGATIONS[2-5]

Introduction

Duffing's equation has no statistical parameters and every solution is uniquely determined by the initial condition. The appearance of statistical properties in the physical phenomena in spite of the perfectly deterministic nature of the equation is caused by the existence of noise in the real systems as well as in the global structure of the solutions. A bundle of solutions representing the chaotically transitional process appears in certain domains of the system parameters. The detalis of these stochastic regions have not been discussed as yet.

Strange Attractor

We have emphasized that the strange attractor is identical with a closure of unstable manifolds of a saddle of the diffeomorphism f_λ. Also, it is defined by the asymptotically stable, invariant, closed set of f_λ containing infinitely many unstable minimal sets connected to one another by the influence of noise in the real system. A decomposition of strange attractors is still open.

It seems that the structure of a strange attractor is unstable in the Andronov-Pontryagin sense. However, the experimental results show that the closure of unstable manifolds always comes out corresponding to the chaotically transitional process and is not affected by the small perturbations of the real system. Judging from these facts, the strange attractor will have structural stability in some sense.

Time Evolution Characteristics

The mean value $m_X(t)$ of the chaotically transitional process $\{X(t)\}$ is a periodic function of t with period 2π. The chaotically transitional process can be regarded as a periodic nonstationary process.

Roughly speaking, though stochastic properties arise from the noise in the real system, the average power spectrum of the process is characterized by the structure of the bundle of solutions alone. The other statistical characteristics of the processes may depend, of course, on the nature of noise in the real system. However, no attempt has been made to determine what that effect is.

Transition of the Processes

When the system parameters are varied between the deterministic and the stochastic regions, strange attractors usually develop from periodic points, producing periodic points of successive twice orders of the original order. Let us call this process

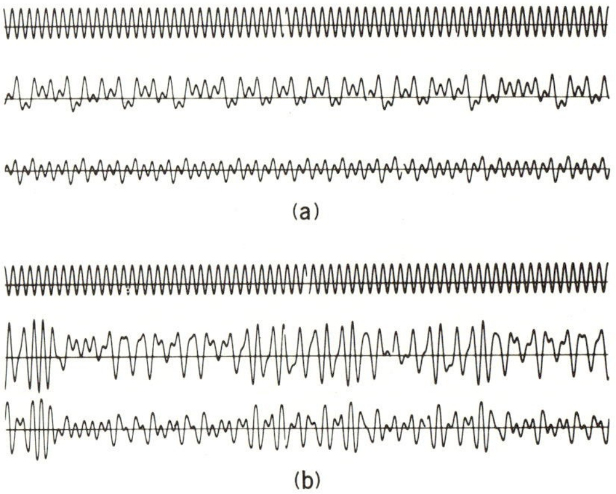

FIGURE 1. Waveforms of the external force and the chaotically transitional processes $\{X(t)\}$ and $\{Y(t)\}$. (a) $k = 0.05$, $B_0 = 0.030$, and $B_1 = 0.16$. (b) $k = 0.05$, $B_0 = 0.045$, and $B_1 = 0.16$.

SI branching because stable points change into inversely unstable points. It is frequently observed that stable periodic points disappear through coalescence with directly unstable periodic points (SD coalescence or SD extinction) and then strange attractors take their place. Strange attractors commonly change into periodic points by tracing SI branching inversely. It sometimes occurs that strange attractors disappear at the spot where the transition chains are formed. That is, as soon as the unstable manifolds composing strange attractors touch another stable manifold, strange attractors cannot generally exist. In any case, the onset and the limit of chaotically transitional processes are closely related to the global structure of the solutions and to the magnitude of uncertain factors in the real system.

EXPERIMENTAL RESULTS ON THE CHAOTICALLY TRANSITIONAL PROCESSES

In this section, we first show representative examples of the chaotically transitional processes exhibited by (2) with two specified values of the system parameters. We then study the transition of the processes by varying one of the system parameters. Finally, we discuss experimental results.

Representative Examples

As is well known, the subharmonic oscillation of order $\frac{1}{2}$ is likely to occur in the system exhibited by Duffing's equation.[6] In this section, let us give two representative examples of the chaotically transitional processes developed from $\frac{1}{2}$ harmonic oscillation. The system parameters used are as follows:

a. $k = 0.05$, $B_0 = 0.030$, and $B_1 = 0.16$,

b. $k = 0.05$, $B_0 = 0.045$, and $B_1 = 0.16$.

(10)

Waveforms and Trajectories

FIGURE 1 shows the waveforms of the external periodic force and the resulting chaotically transitional processes $\{X(t)\}$ and $\{Y(t)\}$. The xy trajectories are given in FIGURE 2. These are the realizations of the chaotically transitional processes.

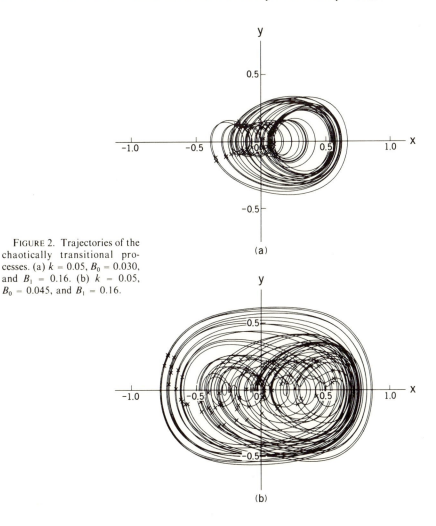

FIGURE 2. Trajectories of the chaotically transitional processes. (a) $k = 0.05$, $B_0 = 0.030$, and $B_1 = 0.16$. (b) $k = 0.05$, $B_0 = 0.045$, and $B_1 = 0.16$.

(a)

(b)

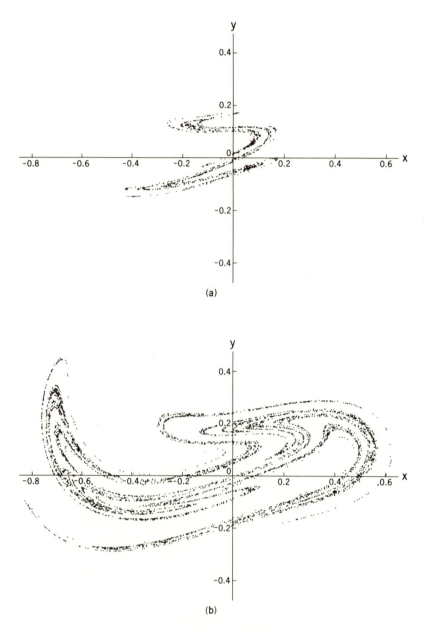

FIGURE 3. Strange attractors of the chaotically transitional processes. (a) $k = 0.05$, $B_0 = 0.030$, and $B_1 = 0.16$. (b) $k = 0.05$, $B_0 = 0.045$, and $B_1 = 0.16$.

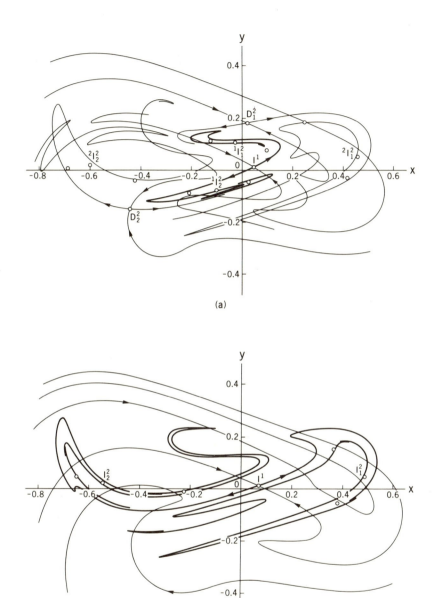

FIGURE 4. Global phase plane structures of the diffeomorphism f_λ. (a) $\lambda = (0.05, 0.030, 0.16)$. (b) $\lambda = (0.05, 0.045, 0.16)$.

Strange Attractors and Global Phase Plane Structures of f_λ

FIGURES 3 and 4 show the strange attractors and the global phase plane structures of the diffeomorphism f_λ. In FIGURE 4, the symbols $^iD_j^n$ and $^iI_j^n$ indicate the ith group of saddles of order n and the subscript j represents the order of the successive movements of the images under f_λ. The symbols D and I mean directly and inversely unstable types, respectively. The unlabeled circles in FIGURE 4 are all inversely unstable 4-periodic points.

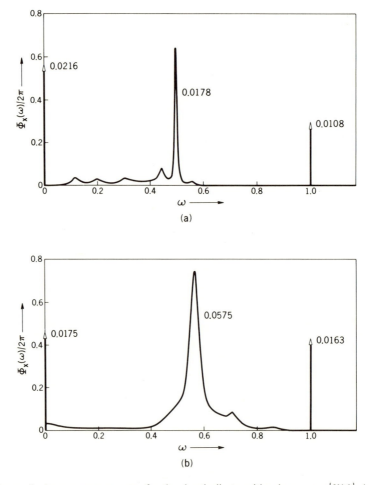

FIGURE 5. Average power spectra for the chaotically transitional processes $\{X(t)\}$. (a) $k = 0.05$, $B_0 = 0.030$, and $B_1 = 0.16$. (b) $k = 0.05$, $B_0 = 0.045$, and $B_1 = 0.16$.

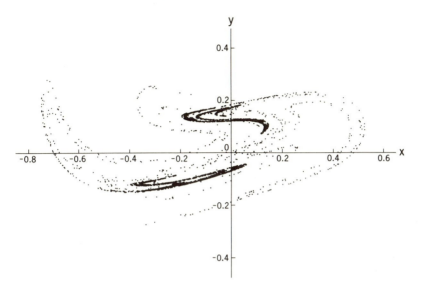

FIGURE 6. A strange attractor has just burst out, the system parameters being $k = 0.05$, $B_0 = 0.035$, and $B_1 = 0.16$.

Time Evolution Characteristics

The mean values of cases a and b are estimated as follows:

$$\text{a. } m_x(t) = 0.147 - 0.207 \cos t + 0.018 \sin t,$$
$$\text{b. } m_x(t) = 0.132 - 0.251 \cos t + 0.044 \sin t. \tag{11}$$

The average power spectra are shown in FIGURE 5. In the figures, line spectra indicate the periodic components and continuous finitenesses show the statistical components. Numerical values in the figures represent the power focused and distributed on those frequencies. The peaks of the statistical components are at $\omega = 0.50$ and $\omega = 0.57$ in cases a and b, respectively.

Transition of the Chaotically Transitional Processes

Let us begin by examining the transition between the above two processes by varying the parameter B_0. When B_0 is increased from the value of case a, the unstable manifold of I^1, $W^u(I^1)$, comes into contact with the stable manifold of D_j^2, $W^s(D_j^2)$, and the explosion of the strange attractor occurs. FIGURE 6 shows the state just burst out for the system parameters $k = 0.05$, $B_0 = 0.035$, and $B_1 = 0.16$. In this strange attractor, the difference in image density is observed but, as B_0 increases, it immediately becomes uniform.

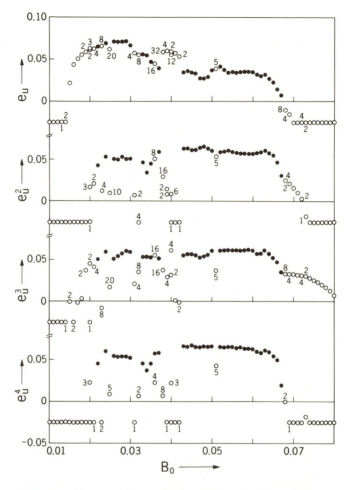

FIGURE 7. Exponentlike quantities e_u, e_u^2, e_u^3, and e_u^4 under the variation of the system parameter B_0.

In order to survey every aspect of the transition, the exponentlike quantities e_u, e_u^2, e_u^3, and e_u^4 are estimated and plotted in FIGURE 7. They are calculated both by increasing and by decreasing the parameter B_0 at an interval of 0.001. In the figure, the chaotically transitional process and the deterministic periodic process are differentiated. The former is marked ● and the latter ○. The number attached to the mark ○ indicates the order of the periodic point. The order of the periodic points without assigned numbers is the same as that of the neighboring one.

Independently of the observable phenomena in the real system, some of the exponentlike quantitites of the deterministic periodic processes take a positive value. This is due to the definition of the exponentlike quantities. However, if the order of the exponentlike quantities is the same as that of the periodic points, then the exponentlike quantities agree with the characteristic exponents of the periodic solution.

Discussion

In this section, we discuss the experimental results obtained in the preceding sections.

1. We see, in FIGURE 4a, that one of the branches of the unstable manifold $W^u(D_j^2)$ constitutes a homoclinic cycle, but the other does not. Let us call this structure partially homoclinic. On the other hand, both branches of $W^u(I^1)$ are homoclinic. Let us call this case totally homoclinic. Furthermore, the saddle D_j^2 is chained to I^1 through transition chain $W^u(D_j^2) \cup W^s(I^1)$, whereas the saddle I^1 is not chained to D_j^2. Therefore, the strange attractor of this case is composed only of the unstable manifold of I^1.

2. The explosion of a strange attractor occurs at the instant when I^1 is chained to D_j^2, that is, when $W^u(I^1)$ touches $W^s(D_j^2)$. The difference in image density in the strange attractor just burst out show the outline of the transition probability density distribution of the chaotically transitional process. Simultaneously with the explosion, continuous finite components of the power spectrum protrude at some frequencies.

3. As B_0 increases, the periodic points $^1I_j^2$ of FIGURE 4a change into completely stable periodic points $^1S_j^2$ at some value of B_0 between 0.039 and 0.040. The points $^1S_j^2$ and D_j^2 of FIGURE 4a then approach each other and, finally, cease to exist through coalescence (SD extinction) at some value of B_0 between 0.042 and 0.043. This gives the outline of the changes of the global phase plane structures from case a to case b.

The points $^1I_j^2$ of FIGURE 4a become stable below 0.021 and the points I_j^2 of FIGURE 4b become stable above 0.073. Needless to say, these transitions are the SI branching.

4. Subharmonic oscillations of various orders appear in the stochastic region. Among them, different from those of order 2^n, appearances of the subharmonic oscillations of orders 3, 5, 12, and 20 are interesting. Though we have not examined this in detail, it seems that the transition from strange attractors to periodic points is due to the transition chain. Let us call this type of transition "extinction of strange attractors." The transitions from periodic points to strange attractors are described above under the heading "Transition of the Processes."

5. We see in FIGURE 7 that, in the intervals of B_0, 0.026–0.030 and 0.043–0.065, the variations of the exponentlike quantities e_u^j for the chaotically transitional processes tend to be uniform as j increases. However, this situation is not present in the interval 0.033–0.037 and the explosion takes place in the interval.

Almost all exponentlike quantities, e_u^j, for the periodic points have an inclination to reduce as j increases. This tendency will be utilized for discriminating a deterministic process from a statistical process.

CONCLUSION

As in our previous reports, experimental results have been introduced relating to the chaotically transitional processes exhibited by Duffing's equation. In particular, the explosion of strange attractors and the appearances of the deterministic periodic points of various orders have been obtained. Furthermore, the whole aspect of the transition is clarified by estimating the exponentlike quantities.

Experimental study using computers produces valuable information about the solutions of Duffing's equation. However, the unsolved problems mentioned above are not yet settled.

ACKNOWLEDGMENTS

The author wishes to thank Professor Chikasa Uenosono of Kyoto University, the President of the Institute of Electrical Engineers of Japan, for his constant support and encouragement during the preparation of this paper.

Professor Joseph Ford of the Georgia Institute of Technology made many useful comments when he visited Kyoto. I greatly appreciate his generous advice in this matter.

This work has been carried out in part under the Collaborating Research Program at the Institute of Plasma Physics of Nagoya University. The author wishes to express his sincere thanks to the staff of the Institute.

REFERENCES

1. JORNA, S., Ed. 1978. Topics in nonlinear dynamics: A tribute to Sir Edward Bullard, AIP Conference Proceedings,Vol. 46. American Institute of Physics. New York.
2. UEDA, Y. et al. 1973. Computer simulation of nonlinear ordinary differential equations and nonperiodic oscillations. Trans. Inst. Electr. Commun. Eng. Jpn. **56A:** 218–25.
3. UEDA, Y. 1978. Random phenomena resulting from nonlinearity—In the system described by Duffing's equation. Trans. Inst. Electr. Eng. Jpn. **98A:** 167–73.
4. UEDA, Y. 1979. Randomly transitional phenomena in the system governed by Duffing's equation. J. Stat. Phys. **20:** 181–96.
5. UEDA, Y. 1979. Steady motions exhibited by Duffing's equation. *In* Engineering Foundation Conf. on New Approaches to Nonlinear Problems in Dynamics, Monterey, California, Dec. 9–14, 1979.
6. HAYASHI, C. 1964. Nonlinear Oscillations in Physical Systems. McGraw-Hill Book Co. New York.

Selection 6

Chaos, Solitons & Fractals Vol. 1. No. 3. pp.199–231. 1991
Printed in Great Britain

0960-0779/91$3.00 + 00
Pergamon Press plc

Survey of Regular and Chaotic Phenomena in the Forced Duffing Oscillator

YOSHISUKE UEDA

Department of Electrical Engineering, Kyoto University, Kyoto 606, Japan

(*Received* 11 *March* 1991)

Abstract — The periodically forced Duffing oscillator

$$\ddot{x} + k\dot{x} + x^3 = B \cos t$$

exhibits a wide variety of interesting phenomena which are fundamental to the behavior of nonlinear dynamical systems, such as regular and chaotic motions, coexisting attractors, regular and fractal basin boundaries, and local and global bifurcations. Analog and digital simulation experiments have provided a survey of the most significant types of behavior; these experiments are essential to any complete understanding, but the experiments alone are not sufficient, and careful interpretation in terms of the geometric theory of dynamical systems is required. The results of the author's survey, begun over 25 yr ago, are here brought together to give a reasonably complete view of the behavior of this important and prototypical dynamical system.

1. INTRODUCTION

Various fascinating and fundamental phenomena occur in nonlinear systems. One of the most representative and the simplest nonlinear systems may be the damped, periodically forced nonlinear oscillator governed by

$$\frac{d^2x}{dt^2} + k\frac{dx}{dt} + f(x) = e(t) \tag{1}$$

where k is a damping coefficient, $f(x)$ is a nonlinear restoring term and $e(t)$ is a periodic function of period T. This equation, first introduced by Duffing [1], has been studied both theoretically and experimentally by many researchers. From the phenomenological point of view, a steady state governed by Duffing's equation (1) may be a periodic motion, the fundamental period of which is either the period T of the external force, or its integral multiple. In more general dynamical systems, a steady state may be an almost periodic motion; however in the case of equation (1), the positive damping k eliminates this possibility [2]. Therefore, for the system under consideration, the regular motions are the periodic steady states. Regular motions have been extensively studied for more than 50 years. However, due to the completely deterministic nature of the equation, the possibility of chaotic motions was overlooked for a long time. The characteristic property of chaotic motion is that its long-term behavior cannot be reproduced in repeated trials from apparently identical initial condition. This contrasts dramatically with the perfect short-term predictability which is guaranteed by the deterministic nature of equation (1).

Even until the beginning of the 1970s, a prejudice existed that there can occur only two kinds of steady states in the second-order nonautonomous periodic systems, that is, periodic and almost periodic motions. A similar prejudice existed among physicists who conjectured that fluid turbulence is a complex form of almost periodic motion. This belief

149

was sharply challenged in 1971 by Ruelle and Takens, who suggested that irregular motions governed by strange attractors are a more likely explanation of turbulence [3]. In fact, the author had already observed an abundance of irregular behavior in second-order systems by this time, reported in [4]; this work achieved wider recognition through Professor Ruelle's articles in *La Recherche* and *The Mathematical Intelligencer* [5, 6], and later in the book of Thompson and Stewart [2].

The following Sections 2–5 contain some mathematical preliminaries [2, 7–16]; Section 2 is essential, but on first reading, Sections 3–5 may be skipped. Sections 6–8 present the results of analog and digital simulations and their interpretation.

2. STROBOSCOPIC OBSERVATION OF THE PHENOMENON: A BRIEF INTRODUCTION TO DISCRETE DYNAMICAL SYSTEMS THEORY [2, 8, 11–17]

Before entering into particular results for the Duffing oscillator, let us briefly explain the fundamental concepts of discrete dynamical systems theory in relation to nonlinear differential equations of the second order.

The equation (1) is a particular case of a nonautonomous periodic system,

$$\frac{dx}{dt} = X(t, x, y), \qquad \frac{dy}{dt} = Y(t, x, y) \tag{2}$$

where $X(t, x, y)$ and $Y(t, x, y)$ are both periodic in t with period T. Here sufficient continuity properties of X and Y are assumed to guarantee the existence and the uniqueness of solutions for any initial condition and for all $t \geqq t_0$.

A solution of equation (2) determines a motion of a representative point on the xy plane. Let us consider the solution

$$x = x(t; t_0, x_0, y_0), \qquad y = y(t; t_0, x_0, y_0) \tag{3}$$

of equation (2) which when $t = t_0$ is at the arbitrary point $p_0(x_0, y_0)$ of the xy plane. The solution (3) describes a curve in the txy space, and the projection of this solution curve on the xy plane represents the phase-plane trajectory of the motion starting from p_0 at $t = t_0$. Let us focus our attention on the location of the representative point $p_n(x_n, y_n)$ at the instant $t = t_0 + nT$, n being $0, 1, 2, \ldots$. An infinite point sequence

$$\{p_0, p_1, p_2, \ldots\} \tag{4}$$

where $x_n = x(t_0 + nT; t_0, x_0, y_0)$, $y_n = y(t_0 + nT; t_0, x_0, y_0)$ is called a positive half-sequence or half-orbit of p_0. This half-orbit represents the behavior of the motion starting from p_0, that is, as time proceeds, or as n tends to infinity, the point p_n approaches, through the transient state, the set of points which represents steady final motion. An accumulation point of the positive half-orbit (4) of p_0 is called an ω-limit point and a set of such points is called an ω-limit set of p_0. An α-limit point and an α-limit set of p_0 are also defined, with reference to a negative half-orbit with $n = 0, -1, -2, \ldots,$. An orbit is a negative half orbit to p_0 plus the positive half-orbit from p_0. An orbit together with its α- and ω-limit sets is called a complete group.

This stroboscopic observation is illustrated in Fig. 1 by putting $t_0 = 0$. The constant t_0 represents the phase of the stroboscopic observation and it can be any chosen value between 0 and T. The choice of t_0 may change the locations of stroboscopic points of the orbit but does not alter their topological structure. Also it should be noted that due to the periodicity of X and Y the translation of the time axis by an arbitrary multiple of T does not alter the situation.

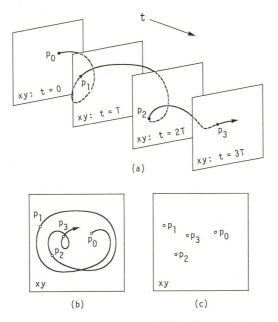

(a)

(b)   (c)

Fig. 1. Schematic illustration of stroboscopic observation. (a) Solution curve in the txy space; (b) trajectory projected onto the xy plane; (c) stroboscopic positive half-orbit in the xy plane.

This stroboscopic observation of the phenomenon can be stated in terms of discrete dynamical systems theory. The solution of equation (2) defines a discrete dynamical system on $\mathbf{R}^2$, or a mapping of the xy plane into itself,

$$f_\lambda: \quad \left. \begin{aligned} \mathbf{R}^2 &\to \mathbf{R}^2 \\ p_0 &\mapsto p_1 \end{aligned} \right\} \tag{5}$$

where $p_1 = f_\lambda(p_0)$ is an image of p_0 under the mapping and λ denotes a set of parameters contained in X and Y of equation (2). We also write the inverse mapping by $p_0 = f_\lambda^{-1}(p_1)$, and nth iterations of the mapping by $p_n = f_\lambda^n(p_0)$. From the properties of the solutions of differential equations, it is known that the mapping (5) is a homeomorphism, that is, a one-to-one continuous mapping with a continuous inverse. Under sufficient smoothness assumptions, this mapping is a diffeomorphism, that is, the mapping and its inverse have continuous derivatives. Finally, the mapping (5) is always orientation-preserving.

By applying the mapping thus introduced to investigate the behaviour of the solution curves in the txy space, we have only to study the successive images of initial points in the xy plane, or the discrete dynamical system in the xy plane into itself. If a solution $(x(t), y(t))$ has period T, then the point $p_0(x(0), y(0))$ is a fixed point of the mapping f_λ. This situation is illustrated in Fig. 2. If a solution has period mT, that is, a solution of period mT but not of period less than mT, the points $p_1, p_2, \ldots, p_m$ are all fixed points of the mapping f_λ^m. Each point is called an m-periodic point, and the totality of these points is called an m-periodic group. This situation is illustrated in Fig. 3 for $m = 2$.

By identifying $t = 0$ with $t = T$, the equation (2) can be transformed to a phase space which is the Cartesian product of the xy plane with the circle representing periodic time.

151

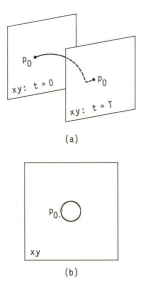

(a)

(b)

Fig. 2. Periodic solution with period T and fixed point p_0.

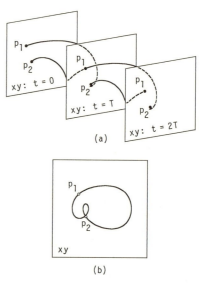

(a)

(b)

Fig. 3. Periodic solution with period $2T$ and two-periodic points p_1 and p_2.

This shows that the stroboscopic map f_λ is an example of a Poincaré map, and the stroboscopic orbit is a Poincaré section of the solution curve in txy space.

3. MAXIMAL BOUNDED INVARIANT SET AND CENTRAL MOTIONS [8, 17–19]

Under rather general conditions satisfied in practice, there exist simple closed curves in the xy plane such that a solution of equation (2) can intersect any one of these curves only by crossing it from the domain exterior to the curve into the domain interior to the curve. Moreover, a curve with this property can be constructed in such a manner that it passes through any point in the xy plane sufficiently remote from the origin and encircles the origin of the xy plane. Therefore a curve Γ_0 can be found such that all solutions starting outside Γ_0 must eventually pass through Γ_0 to the interior and remain inside for all subsequent time.

If Γ_0 denotes a simple closed curve of this type, it follows that under iterations of the mapping f_λ all points exterior to Γ_0 are transferred into interior points. In practice we may choose Γ_0 such that the interior Δ_0 of Γ_0 is mapped under one iteration to the domain Δ_1 interior to $\Gamma_1 = f_\lambda(\Gamma_0)$. If $f_\lambda^n(\Gamma_0)$ is denoted by Γ_n and the closed domain bounded by Γ_n in the xy plane by Δ_n, then Δ_n lies in the domain interior to Δ_{n-1}, that is, $\Delta_n \subset \Delta_{n-1}$ holds.

The closed set

$$\Delta = \bigcap_{n=0}^{\infty} \Delta_n \qquad (6)$$

is called the maximal bounded invariant set.

A similar construction was given by Levinson for a class of systems (2) which are dissipative at large displacements, or class D; he showed that the maximal bounded invariant set so constructed is closed, connected, and has the property that images of a point which is not contained in Δ tend to Δ under iterations of the mapping f_λ. This invariant set Δ contains all the dynamics of the steady states in the system (2), however, it contains points representing not only steady states but also some transient states.

Let us introduce the concepts of non-wandering set and central motions introduced by G. D. Birkhoff. It was Birkhoff who first investigated the comprehensive mathematical description of steady states in nonlinear systems. Consider an arbitrary connected region σ on the xy plane. It may happen that σ is intersected by none of its images

$$\ldots, \sigma_{-2}, \sigma_{-1}, \sigma_1, \sigma_2, \ldots \tag{7}$$

where $\sigma_n = f_\lambda^n(\sigma)$, in which case σ is called a wandering region and its points wandering points. Obviously, wandering points represent transient states; let us denote the totality of wandering points of the xy plane by W^1. A point in the xy plane which is contained in no wandering region is called a non-wandering point. The totality of non-wandering points of the xy plane constitutes a non-empty closed invariant set M^1 towards which all other points of Δ tend asymptotically on indefinite iteration of f_λ or f_λ^{-1}.

Let us suppose now that M^1 is not identical with Δ, and let us take the set M^1 as fundamental instead of Δ. A connected region which contains points of M^1 will be called wandering with respect to M^1 if the set $\sigma \cap M^1$ of points common to σ and M^1 is intersected by none of its images under powers of f_λ or f_λ^{-1}. The points of M^1 which are contained in such a region are called wandering with respect to M^1, and their totality is denoted by W^2. The set $M^2 = M^1 - W^2$ consists of the points which are non-wandering with respect to M^1. In the case where $M^1 = M^2$, M^1 is called non-wandering with respect to itself.

In principle this construction can be repeated; Birkhoff composed a sequence

$$M^1, M^2, M^3, \ldots \tag{8}$$

where each set is non-empty and a proper subset of all those preceding it. He proved that this sequence terminates with $M^r = M^{r+1} = \ldots$ which therefore must be non-wandering with respect to itself. The points of M^r are called central points, and motions started from central points are called central motions.

In the 1960s, C. Pugh proved a remarkable theorem, the so-called closing lemma, which has a corollary that in a generic system the periodic orbits are dense in the non-wandering set M^1 [20]. Hence every point in M^1 is non-wandering with respect to M^1, and M^1 is equal to the set of central motions. Although numerical studies by their nature cannot confirm this mathematical theorem, we may say that extensive numerical experience reveals no contradiction to Pugh's results.

Here let us decompose a set of central points. A point which is both an α- and ω-limit point is called pseudo-recurrent. The characteristic property of such a point is that it returns infinitely often into an arbitrary small neighborhood of itself under indefinite iteration of f_λ as well as f_λ^{-1}. Birkhoff proved that the set of central points consists of the pseudo-recurrent points together with their derived set, a central point is either a pseudo-recurrent point or an accumulation point of pseudo-recurrent points. The complete group of a pseudo-recurrent point is called a quasi-minimal set.

If a non-empty invariant closed set in a quasi-minimal set is identical with the quasi-minimal set itself, the quasi-minimal set is called a minimal set, and points of a minimal set are called recurrent points. Obviously, an orbit of a recurrent point, or the minimal set itself represents a single final motion or the smallest unit of a theoretical or an ideal steady state.

153

4. FIXED AND PERIODIC POINTS AND RELATED PROPERTIES [17, 21]

Let us consider the classification of fixed and periodic points according to the type and the number of these points contained in the maximal bounded invariant set Δ. These are the simplest type of minimal sets among the solutions of equation (2). The behavior of successive images in the neighborhood of a fixed or a periodic point determines the stability of the associated periodic solution. Let $p_0(x_0, y_0)$ be a fixed point and $q_0(x_0 + \xi_0, y_0 + \eta_0)$ be a neighboring point of p_0. Let $q_1(x_0 + \xi_1, y_0 + \eta_1)$ be an image of q_0 under the mapping f_λ, then the following relation results

$$
\left.
\begin{aligned}
x_0 + \xi_1 &= x(t_0 + T; t_0, x_0 + \xi_0, y_0 + \eta_0) \\
y_0 + \eta_1 &= y(t_0 + T; t_0, x_0 + \xi_0, y_0 + \eta_0)
\end{aligned}
\right\} . \tag{9}
$$

This situation is illustrated in Fig. 4(a) with $t_0 = 0$. As shown in the figure, when q_0 is varied to encircle p_0, its image q_1 traces out an ellipse around p_0 in the same sense as q_0. For small values of ξ_0 and η_0, ξ_1 and η_1 can be expanded into power series in ξ_0 and η_0,

$$
\left.
\begin{aligned}
\xi_1 &= a\xi_0 + b\eta_0 + \ldots \\
\eta_1 &= c\xi_0 + d\eta_0 + \ldots
\end{aligned}
\right\} \tag{10}
$$

with $a = (\partial x/\partial \xi_0)_0$, $b = (\partial x/\partial \eta_0)_0$, $c = (\partial y/\partial \xi_0)_0$ and $d = (\partial y/\partial \eta_0)_0$ where $(*)_0$ denotes the value of $(*)$ at $\xi_0 = \eta_0 = 0$. The terms not explicitly given in the right-hand side of equation (10) are of degree higher than the first in ξ_0 and η_0. Equation (10) describes the mapping $q_0 \mapsto q_1$ in the neighborhood of the fixed point p_0, and this mapping is characterized by the roots ρ_1 and ρ_2 of the equation,

$$
\begin{vmatrix} a - \rho & b \\ c & d - \rho \end{vmatrix} = 0. \tag{11}
$$

Since the roots ρ_1 and ρ_2 are determined from the quadratic equation, they are either both real or else are conjugate complex. However, in this case, from the general theory of differential equations, their product is always positive, because the following relation holds,

$$
\begin{vmatrix} a & b \\ c & d \end{vmatrix} = \exp\left[\int_{t_0}^{t_0 + T} \left(\frac{\partial X}{\partial x} + \frac{\partial Y}{\partial y} \right)_0 dt \right] \tag{12}
$$

where $(*)_0$ denotes the value of $(*)$ at the periodic solution under consideration.

The fixed point p_0 is called simple if both $|\rho_1|$ and $|\rho_2|$ are different from unity. If one or both $|\rho_1|$ and $|\rho_2|$ is unity, this means that the fixed point is multiple. Levinson classified simple fixed points as follows:

sink or completely stable if $|\rho_1| < 1, |\rho_2| < 1$
source or completely unstable if $|\rho_1| > 1, |\rho_2| > 1$
saddle or directly unstable if $0 < \rho_2 < 1 < \rho_1$
saddle or inversely unstable if $\rho_2 < -1 < \rho_1 < 0$.

The same classification also applies to periodic points.

If a fixed point is a sink, there is a neighborhood of the fixed point which contracts to the point as shown in Fig. 4(a) and (b); all points in this neighborhood tend to the fixed point under repeated application of the mapping f_λ. This implies that as t tends to infinity, the corresponding solutions tend to the periodic solution, so that this periodic solution is asymptotically stable in the sense of Lyapunov. If a fixed point is a source, a neighborhood

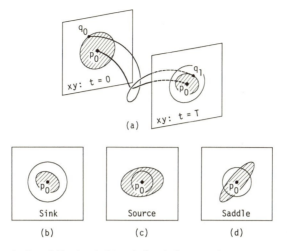

Fig. 4. Behavior of images in the neighborhood of a periodic solution according to stability type of the fixed point. (a) Schematic txy diagram; (b) sink; (c) source; (d) saddle.

of the fixed point expands from the point as shown in Fig. 4(c) and all points in this neighborhood move away from the fixed point under the mapping f_λ. If a fixed point is a saddle, we have the situation of a neighborhood which expands and contracts locally under the mapping f_λ as shown in Fig. 4(d). In this case, there abut at the saddle four invariant curves or branches: two α-branches, whose points converge toward the saddle on iteration of f_λ^{-1}, and two ω-branches, whose points converge toward the saddle on iteration of f_λ. For the saddle, the difference between the directly and inversely unstable cases is illustrated in Fig. 5. Here successive locations of a periodic solution of equation (2) through one period are shown corresponding to a directly unstable fixed point D and an inversely unstable point I. By choosing the strobing angle at 12 different values progressing from 0 to T, we follow the rotation, expansion and contraction of the local α- and ω-branches. Note that for purpose of illustration, the phase plane at $t = T$ is slightly shifted from its position at $t = 0$, so that the final image of D (or I) may be distinguished from the initial image. In each case, the circular dot identifies an expanding α-branch and the square dot a contracting ω-branch. In the case of D, there is one full rotation during the period $0-T$, while in the case of I there is a half-rotation over the period. This is only the simplest case; in general, D might make an integer number of rotations during one period, and I might make a half-integer number of rotations. Only the local linear portion of the α- and ω-branches are shown; larger portions would exhibit curvature due to nonlinearity.

Levinson and Massera have discussed the number of fixed points and periodic points of equation (2) in the xy plane. Let $N(n)$ be the total number of n-periodic points and $C(n)$ the total number of completely stable and completely unstable n-periodic points. Similarly, let $D(n)$ and $I(n)$ be the number of directly unstable and inversely unstable n-periodic points, respectively. When equation (2) has a maximal bounded invariant set and all periodic points are simple, we have the following.

For $n = 1$,

$$\left. \begin{array}{c} C(1) + I(1) = D(1) + 1, \\ N(1) = 2D(1) + 1 \end{array} \right\}. \tag{13}$$

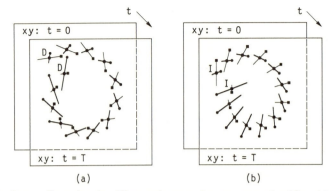

Fig. 5. Schematic diagram illustrating the difference between the two types of saddles. (a) Directly unstable saddle; (b) inversely unstable saddle.

For $n = 2, 4, 6, \ldots$,

$$\left.\begin{array}{l} C(n) + I(n) = D(n) + 2I(n/2), \\ N(n) = 2[D(n) + I(n/2)] \end{array}\right\}. \tag{14}$$

For $n = 3, 5, 7, \ldots$,

$$\left.\begin{array}{l} C(n) + I(n) = D(n), \\ N(n) = 2D(n) \end{array}\right\}. \tag{15}$$

These relations give an independent verification of whether a numerical search for fixed points or periodic points is complete. If the number and type of points observed do not conform to these relations, then undoubtedly there are additional points to be located; but of course the converse need not be true: the relations might be satisfied for a partial set of fixed and periodic points.

After the fixed and periodic points, the next simplest type of minimal set is an invariant closed curve in the stroboscopic or Poincaré section. As noted above, this phenomena cannot occur in a system with uniformly positive damping, so we shall not discuss it further here, and instead refer the reader to [17].

5. DOUBLY ASYMPTOTIC POINTS AND RELATED PROPERTIES [7, 8]

In systems described by autonomous differential equations of the second order, no two trajectories in the xy plane can intersect each other, and they may only approach each other asymptotically at the equilibrium points. However, in nonautonomous periodic systems of the second order, a somewhat different situation occurs: invariant branches of saddle points of the mapping f_λ^n may intersect one another, introducing a complicated structure into the dynamics. Here let us introduce some terminology and concepts defined by Poincaré, which describe this complexity.

We have already explained the α- and ω-branches of the saddles of the mapping f_λ^n. If we consider the totality of α- and ω-branches of fixed or periodic points of all orders in the *xy* plane, it is readily shown that no α- (or ω-) branch can intersect another α- (or ω-) branch. However, an α-branch may intersect an ω-branch, and the points of intersection in such a case are called doubly asymptotic points.

A doubly asymptotic point is called homoclinic if the α- and ω-branches on which it lies issue from the same point or from two points belonging to the same periodic group. A homoclinic point of the former type is called simple. A doubly asymptotic point is called heteroclinic if the α- and ω-branches on which it lies issue from two periodic points, each of them belonging to different periodic groups. Also a doubly asymptotic point is said to be of general type if the two branches which intersect at the doubly asymptotic point cross transversely, that is, are not coincident or merely tangent at the point; in the contrary case the point is of special type. When a mapping has no doubly asymptotic points of special type and no fixed or periodic points of multiple type, the mapping is said to belong to the general analytic case.

Birkhoff proved the following theorems.

In the general analytic case, an arbitrary small neighborhood of a homoclinic point contains infinitely many periodic points.

In the general analytic case, an arbitrary small neighborhood of a homoclinic point contains a homoclinic point of simple type.

In order to illustrate these situations, Fig. 6 shows an example of a homoclinic point of simple type, together with sketches of portions of the α-branch (thick line) and the ω-branch (fine line). The existence of a homoclinic point H implies the existence of additional homoclinic points H_1, H_2, H_3, ..., and indeed an infinite sequence of homoclinic points approaching D along its ω-branch, each point being the image of its predecessor under f_λ. Likewise, successive pre-images H_{-1}, H_{-2} and so on are homoclinic points approaching D along its α-branch. Near D, the stretching and contraction of a neighborhood of H is governed by the local linear approximation of f_λ; the images of such a neighborhood, which are schematically shown by shaded regions, become extremely long and thin under iteration of f_λ near D. An example of a periodic point with images p_1, p_2, p_3, p_4, p_5, p_6 is indicated; additional periodic points of longer period lie nearer to the homoclinic points.

The original terms α- and ω-branch have been retained for reasons of nostalgia; in current parlance, the α-branch is called the unstable manifold, while the ω-branch is called the stable manifold. Another alternative and perhaps more suggestive terminology was proposed by Zeeman, who calls the α-branch the outset and the ω-branch the inset [22].

Up to this point, we have considered a mathematical description of behavior of an ideal dynamical system which is perfectly deterministic, not subject to any noise or disturbance, and described with total precision.

In what follows, we will investigate the behavior of this ideal system through real-world simulations using analog and digital computing devices. This brings unavoidable additional influences such as small amounts of noise, imperfect precision, and numerical errors due to approximate solution algorithms and roundoff. Although it may not be easy to incorporate these effects into rigorous mathematical analysis, these features of real-world simulations have a healthy influence in guiding our attention to the aspects of the invariant set which correspond to robust and stable behavior.

6. COLLECTION OF STEADY STATES IN THE DUFFING OSCILLATOR [23, 24]

As a specific example of equation (2), let us consider the Duffing equation

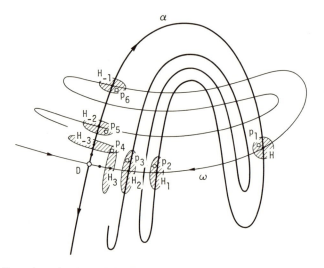

Fig. 6. Schematic illustration of a homoclinic structure, showing homoclinic points and neighboring periodic points.

$$\frac{\mathrm{d}^2 x}{\mathrm{d}t^2} + k \frac{\mathrm{d}x}{\mathrm{d}t} + x^3 = B \cos t \tag{16}$$

or

$$\frac{\mathrm{d}x}{\mathrm{d}t} = y, \qquad \frac{\mathrm{d}y}{\mathrm{d}t} = -ky - x^3 + B \cos t. \tag{17}$$

This equation represents forced oscillations in a variety of applications, with constant positive damping coefficient k, and nonlinear restoring force x^3 representing the simplest form of hardening symmetric spring in a mechanical system, or of magnetic saturation in an electrical circuit with a saturable core inductor [25–27]. In the derivation of such an equation, some approximations and simplifying assumptions are introduced, and small fluctuations and noise are neglected.

One might also consider an apparently more general system with angular forcing frequency ω different from one and coefficient of x^3 different from one. However, such a system can always be brought to the form (16) by appropriately rescaling x and t. Although there are advantages to considering forcing frequency ω as a parameter, we use k and B as the coordinates of parameter space, and do not explicitly consider transformation to other equivalent parameter space coordinates such as ω and B.

The symmetry of equation (17), associated with its invariance under the substitution $-x \to x$, $-y \to y$, $t + \pi \to t$, implies that a periodic trajectory is either symmetric to itself with respect to the origin of the xy plane or it coexists with another periodic trajectory symmetric to it with respect to the origin. Note however that the stroboscopic points of symmetrically related trajectories will not appear to be symmetric unless one of the pair is strobed with phase shift π. Also the positive damping k results in the non-existence of sources and of invariant closed curves representing almost periodic motions; and the area of the maximal bounded invariant set Δ of the mapping f_λ defined by equation (17) is necessarily zero.

In the damped, forced oscillatory system given by equation (16), various types of steady states are observed depending on the system parameters $\lambda = (k, B)$ as well as on the initial conditions. Figure 7 shows the regions on the kB plane in which different steady motions are observed. The regions are obtained by both analog and digital simulations. The roman numerals I, II, II', III and IV characterize periodic motions with period 2π. The fractions m/n ($m = 1, 3, 4, 5, 6, 7, 11$ and $n = 2, 3$) indicate the regions in which subharmonic or ultrasubharmonic motions of order m/n occur. An ultrasubharmonic motion of order m/n is a periodic motion of period $2n\pi$, and whose principal frequency is m/n times the frequency of the external force. In addition to these regular steady states, chaotic motions take place in the shaded regions. In the area hatched by solid lines, chaotic motion occurs uniquely, independent of initial condition; while in the area hatched by broken lines, two different steady states coexist, one being chaotic and the other a regular motion. Which one occurs depends on the initial condition. Ultrasubharmonic motions of higher orders ($n = 4, 5, \ldots,$) can occur naturally in the system, but they exist only in narrow regions and are therefore omitted in Fig. 7. Also some subtle details of region boundaries are not represented.

In order to clarify the meaning of this kB chart, we choose a set of typical parameters $\lambda = (k, B)$ from each region. Their locations are indicated by letters from a to u in Fig. 7. Figure 8 shows the steady states trajectories observed at each of these 21 parameter values. In the figure we can see multiple steady states for certain parameter values. On the trajectories, a $\times$ marks the location of a stroboscopic point at the instant $t = 2i\pi$ (i being integer). Therefore these $\times$ marks show the sinks of the mapping f_λ^n ($n = 1, 2, 3$) in all cases where the attractor is a regular periodic motion, that is, all cases except (k), (l_1) and (o_1).

The three cases (k), (l_1) and (o_1) show chaotic motions in which the trajectories are drawn after the transient states have died away. The remarkable feature of the chaotic steady states is that, however small simulation error may be, the precise long-term trajectory cannot be reproduced in repeated experiments; but nevertheless the same structure always eventually develops. This is seen most clearly by stroboscopic sampling. To see the situation more clearly, only one trajectory was computed for a long time and its steady state stroboscopic orbit is shown in Fig. 9 for these three cases. The orbits thus obtained indicate the presence of chaotic attractors, that is, steady state motions with definite structure and also an aspect of randomness. Figure 10 shows corresponding waveforms obtained by analog simulation, which may be impossible to distinguish from outcomes of a random process.

As shown above multiple attractors are common in the system (17) and indeed there may be regular and chaotic attractors coexisting, for example, for the cases (l) and (o). Here we would like to associate each attractor with the ensemble of all starting points that settle to it; this point set is called its basin of attraction. These basins are separated by ω-branches of some saddle points in the xy plane, and together the basins of all attractors and their separators (basin boundaries) constitute the entire xy plane.

Figure 11 shows attractor-basin phase portraits at successive times differing by phase $\pi/6$, for the parameter values (o). In this case the maximal bounded invariant set consists of the following: a sink (circle), a directly unstable saddle (filled circle) together with its α-branches, and a chaotic attractor. The non-wandering set M^1 is the sink, saddle and chaotic attractor. In the sequence shown, one folding and stretching action is completed, and in the interval from $t = \pi$ to $t = 2\pi$ a second, symmetrically related folding and stretching is accomplished. As will be seen later, the chaotic attractor contains exactly three saddle fixed points, one of directly unstable type and two of inversely unstable type.

Here, let us proceed to the statistical time series analysis for a chaotic attractor. To this end, we regard the chaotic motion $\{X(t)\}$ as a periodic process $\{X_T(t)\}$ with a sufficiently long period T, where T is a multiple of 2π. Note that in this section only, we use T exclusively to stand for this very long observation interval, and not in the previous meaning

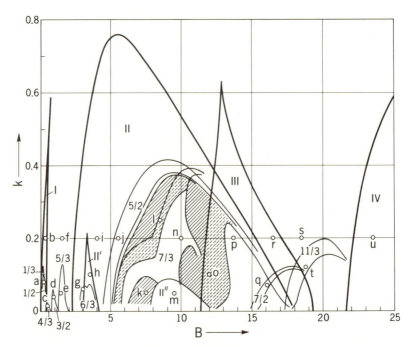

Fig. 7. Regions of different steady states for the system defined by equation (16). (Reproduced with the courtesy of the Society for Industrial and Applied Mathematics [24].)

of the period of the system, which is here taken equal to 2π.

Now suppose $x_T(t)$ to be a periodic function with period T, which coincides with a realization $x_T(t)$ of $\{X_T(t)\}$ in the interval $(-T/2, T/2]$. Then the realization $x_T(t)$ is expanded into Fourier series as

$$x_T(t) = \frac{a_0}{2} + \sum_{m=1}^{\infty} (a_m \cos m\omega_0 t + b_m \sin m\omega_0 t), \qquad \omega_0 = \frac{2\pi}{T} \tag{18}$$

where

$$
\left.
\begin{aligned}
a_m &= \frac{2}{T} \int_{-T/2}^{T/2} x_T(t) \cos m\omega_0 t \, dt \\
b_m &= \frac{2}{T} \int_{-T/2}^{T/2} x_T(t) \sin m\omega_0 t \, dt \\
m &= 0, 1, 2, \ldots
\end{aligned}
\right\} \tag{19}
$$

Fourier coefficients a_m and b_m can not be distinguished from random variables because $x_T(t)$ is a sample function of the process $\{X_T(t)\}$. From these coefficients, the mean value $m_X(t)$ and the average power spectrum $\Phi_X(\omega)$ of the process $\{X(t)\}$ can be estimated as follows.

$$
\begin{aligned}
m_X(t) = \langle X(t) \rangle &= \lim_{T \to \infty} \langle X_T(t) \rangle \\
&\doteqdot \langle X_T(t) \rangle = \left\langle \frac{a_0}{2} \right\rangle + \sum_{m=1}^{\infty} [\langle a_m \rangle \cos m\omega_0 t + \langle b_m \rangle \sin m\omega_0 t] \tag{20}
\end{aligned}
$$

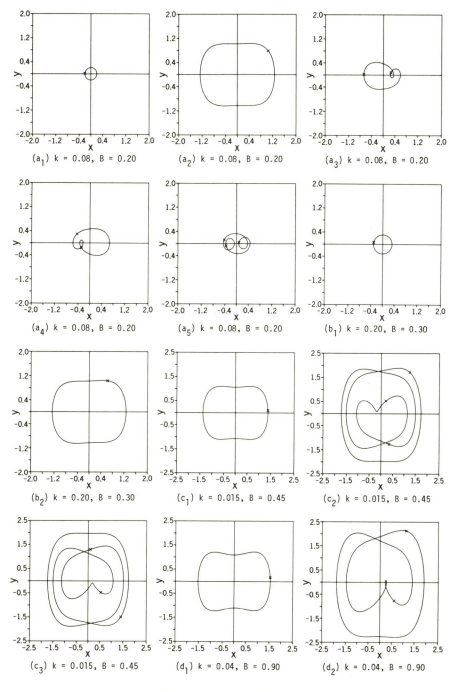

(a_1) k = 0.08, B = 0.20

(a_2) k = 0.08, B = 0.20

(a_3) k = 0.08, B = 0.20

(a_4) k = 0.08, B = 0.20

(a_5) k = 0.08, B = 0.20

(b_1) k = 0.20, B = 0.30

(b_2) k = 0.20, B = 0.30

(c_1) k = 0.015, B = 0.45

(c_2) k = 0.015, B = 0.45

(c_3) k = 0.015, B = 0.45

(d_1) k = 0.04, B = 0.90

(d_2) k = 0.04, B = 0.90

Fig. 8. continued on p. 212.

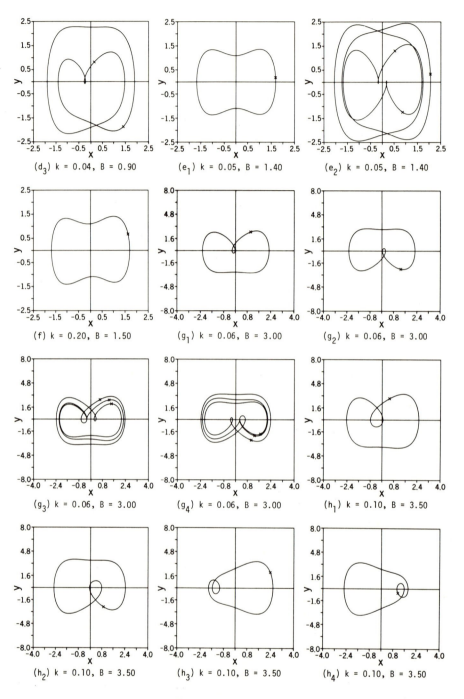

(d_3) k = 0.04, B = 0.90

(e_1) k = 0.05, B = 1.40

(e_2) k = 0.05, B = 1.40

(f) k = 0.20, B = 1.50

(g_1) k = 0.06, B = 3.00

(g_2) k = 0.06, B = 3.00

(g_3) k = 0.06, B = 3.00

(g_4) k = 0.06, B = 3.00

(h_1) k = 0.10, B = 3.50

(h_2) k = 0.10, B = 3.50

(h_3) k = 0.10, B = 3.50

(h_4) k = 0.10, B = 3.50

Fig. 8. continued on p. 213.

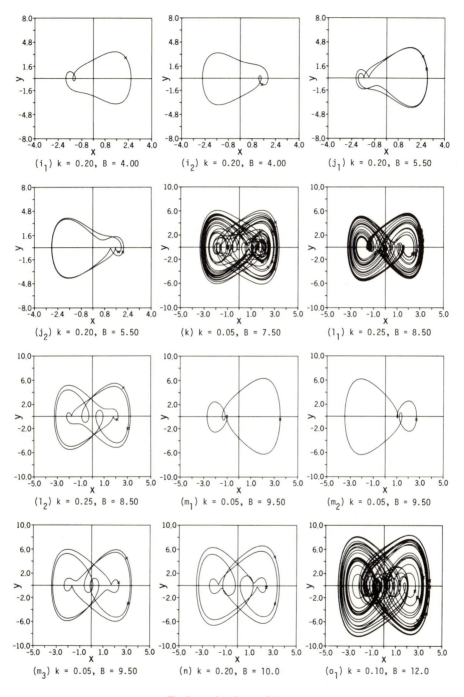

Fig. 8. continued on p. 214.

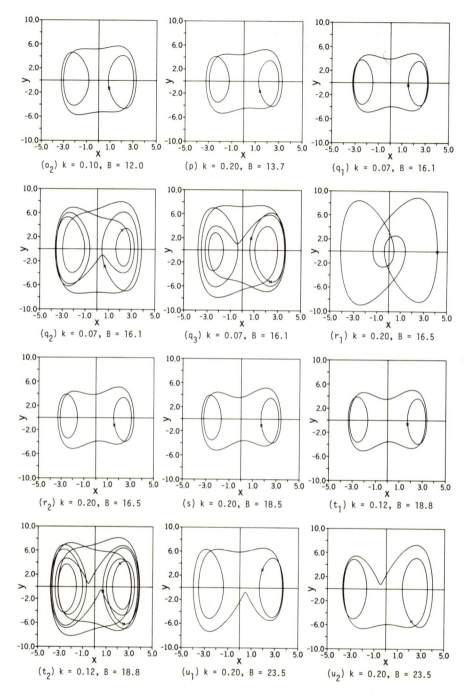

Fig. 8. Trajectories of various types of steady motion. (Reproduced with the courtesy of the Society for Industrial and Applied Mathematics [24].)

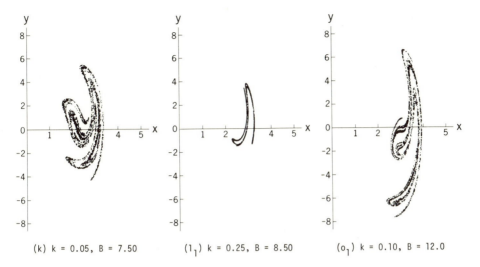

(k) k = 0.05, B = 7.50   (1₁) k = 0.25, B = 8.50   (o₁) k = 0.10, B = 12.0

Fig. 9. Chaotic attractors for three representative sets of parameter values. (Reproduced with the courtesy of the Society for Industrial and Applied Mathematics [24].)

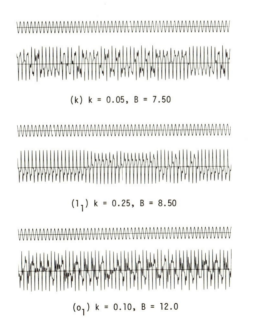

(k) k = 0.05, B = 7.50

(1₁) k = 0.25, B = 8.50

(o₁) k = 0.10, B = 12.0

Fig. 10. Waveforms corresponding to the chaotic attractors of Fig. 9 obtained by analog simulation. (Reproduced with the courtesy of the Society for Industrial and Applied Mathematics [24].)

$$\Phi_X(\omega) = \lim_{T \to \infty} \left\langle \frac{1}{T} \left| \int_{-T/2}^{T/2} x_T(t) e^{-i\omega t} \, dt \right|^2 \right\rangle$$

$$\doteq \Phi_X(m\omega_0) = \frac{2\pi}{\omega_0} \left\langle \frac{1}{4} (a_m^2 + b_m^2) \right\rangle, \ \omega_0 = \frac{2\pi}{T}. \tag{21}$$

YOSHISUKE UEDA

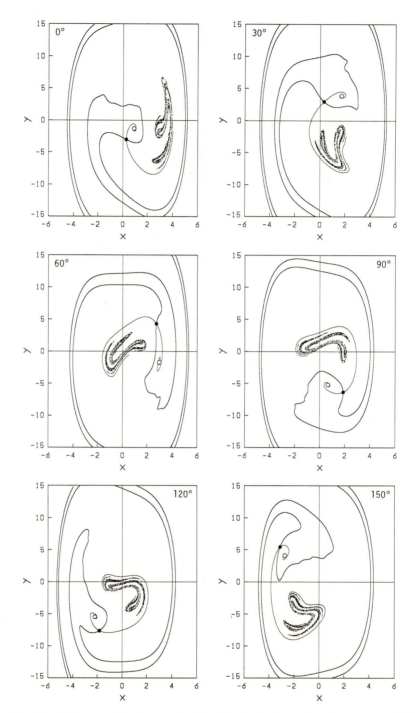

Fig. 11. Attractor-basin phase portraits at successive times differing by phase $\pi/6$ at the parameters (o) of Fig. 7 showing folding and stretching action.

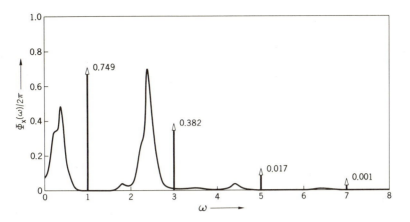

Fig. 12. Average power spectrum of the chaotic motion corresponding to the chaotic attractor of Fig. 9(o_1). (Reproduced with the courtesy of the Institute of Electrical Engineers of Japan [23].)

The ensemble average can be calculated by regarding successive waveforms in the intervals $((n - 1/2)T, (n + 1/2)T]$, $(n = 0, 1, 2, \ldots, N_S)$ as sample functions of $\{X_T(t)\}$.

Let us give an example thus estimated for the representative case of the system parameters (o), that is, $\lambda = (k, B) = (0.1, 12.0)$. The mean value of $\{X(t)\}$ was computed by Fast Fourier Transforms of numerical solutions over observation intervals $T = 2\pi \times 2^{10}$ with 100 samples, and was found to be

$$
\begin{aligned}
m_X(t) = {} & 1.72 \cos t + 0.22 \sin t \\
& + 1.21 \cos 3t - 0.26 \sin 3t \\
& + 0.25 \cos 5t - 0.06 \sin 5t \\
& + 0.07 \cos 7t - 0.02 \sin 7t \\
& + 0.02 \cos 9t - 0.01 \sin 9t.
\end{aligned}
\tag{22}
$$

The mean value is found to be a periodic function. This indicates that the process $\{X(t)\}$ is a periodic non-stationary process. Figure 12 shows the average power spectrum estimated by using equation (21). In the figure, line spectra at $\omega = 1, 3, 5, \ldots$, indicate the periodic components of the mean value as given by equation (22), and numerical values attached to line spectra represent the power concentrated at those frequencies. Besides the line spectra, there are continuous power spectra representing the chaotic component. The average power of this process $\{X(t)\}$ is given by

$$
\lim_{T \to \infty} \frac{1}{T} \int_{-T/2}^{T/2} \langle X_T^2(t) \rangle \, \mathrm{d}t \doteqdot \frac{1}{T} \int_{-T/2}^{T/2} \langle X_T^2(t) \rangle \, \mathrm{d}t = 3.08.
\tag{23}
$$

By adding noise in the numerical integration scheme, we have also confirmed that chaotic attractors as well as averages and spectra of the corresponding chaotic motions are insensitive to numerical error or noise in the computer experiment. Therefore, it was conjectured that the average power spectrum of the process is a characteristic of the global structure of the chaotic attractor, independent of the nature of small uncertain factors or microfluctuations in the actual systems.

7. COLLECTION OF ATTRACTOR-BASIN PHASE PORTRAITS [28]

A typical example of an attractor-basin phase portrait was shown in the preceding section as the stroboscopic Poincaré section at various phases. In these portraits, basin boundaries were given by ω-branches of saddles of the mapping f_λ in the xy plane. These boundaries were located by tracing ω-branches from the saddles by reversing time in the simulation. For the example given in Fig. 11, ω-branches were rather simple and smooth, hence they were located with fairly good accuracy. However, as will be seen later, for the case of fractal basin boundaries, it becomes very difficult with this method of analysis to locate the basin boundaries, because ω-branches become infinitely stretched and folded by homoclinic structure, therefore we cannot avoid considerable error in locating fine details of tangled basin boundaries by numerical experiment. In these situations, the full complexity of basin boundaries must be identified by exhaustive trial of various starting conditions.

In order to supplement the preceding collection of steady states, let us here survey attractor-basin phase portraits for additional representative sets of system parameters $\lambda = (k, B)$. Figures 13 and 14 illustrate the difference between smooth and fractal basin boundaries. In both cases, there are two attractors (sinks) represented by small circles; the shaded regions show the basins of attraction of the resonant motions, while the blank regions represent non-resonant ones. There exist directly unstable fixed points (saddles), represented by filled circles, in their basin boundaries. In computing these figures, and all attractor-basin portraits hereafter, a fourth-order Runge–Kutta–Gill scheme with fixed step size was used with single precision: initial conditions were chosen on a uniform grid of points, integrations being continued until final behavior was confirmed for each grid point. The integration step sizes and the numbers of grid points are given in each figure caption as well as the values of the system parameters. In Fig. 14, a sequence of successive enlargements of smaller and smaller regions of the xy plane clearly shows the Cantor set structure of the basin boundaries. There is no end to this enlargement sequence, and similar geometric structure continues infinitely. This property of the basin boundary is called self-similar or fractal. The fractal nature of the basin boundary originates from the homoclinic structure of the invariant branches of the saddle, that is, the α-branch which tends toward the non-resonant attractor (not shown in the figure) crosses the ω-branch which tends toward the saddle from the right side. Some of the homoclinic structure is illustrated in Fig. 1(b) of [29], Fig. 6 of [30] and Fig. 4 of [31]. Also omitted from Fig. 14 are extremely narrow basins of a pair of extremely small two-periodic chaotic attractors which were detected inside the blank region; these small attractors exist only in a very narrow range of parameter values. Note that the characteristic property of a fractal basin boundary is that transient behavior started from the boundary is chaotic and consequently the final steady motion becomes indeterminate.

Before presenting the other attractor-basin portraits, we must here explain our symbols for the fixed and periodic points. We use the following symbols: ($\bigcirc$) for sink or completely stable fixed point; ($\bullet$) for saddle or directly unstable fixed point; ($\blacksquare$) for saddle or inversely unstable fixed point; ($\times$, $+$) for n-periodic point ($n = 2, 3$). The periodic points inside the basins are sinks and ones in the basin boundaries are saddles. As was shown in Fig. 8, multiple periodic groups (attractors) are common; periodic points belonging to the same group are marked by the same symbol. However, in order to avoid complexity, basins of different but related periodic points are not always distinguished. The order of successive movement of images under the mapping f_λ among these basins can be easily seen from the corresponding cases in Fig. 8. Figure 15 shows an attractor-basin phase portrait corresponding to the point (c) on the kB chart of Fig. 7, and final motions are the fundamental harmonic and ultrasubharmonics of order 4/3. Figure 16

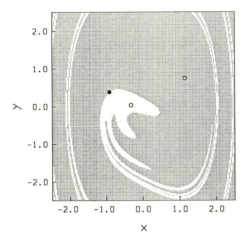

Fig. 13. Attractor-basin phase portrait with integration step size $2\pi/60$ and 201×201 grid of initial conditions for $k = 0.1$, $B = 0.3$. (Reproduced with the courtesy of the European Conference on Circuit Theory and Design [28].)

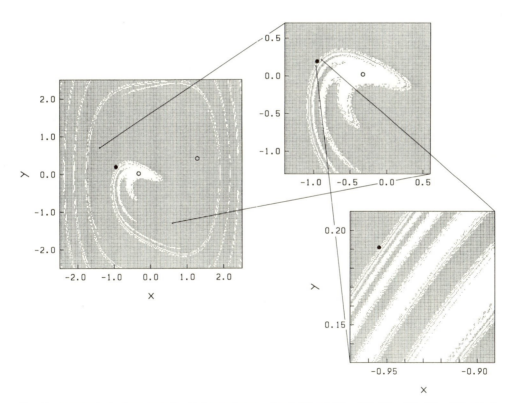

Fig. 14. Attractor-basin phase portrait with integration step size $2\pi/60$ and 201×201, 161×161, 161×161 grid of initial conditions for $k = 0.05$, $B = 0.3$. (Reproduced with the courtesy of the European Conference on Circuit Theory and Design [28].)

corresponds to the point (d), and ultrasubharmonics of order $3/2$. Figure 17 corresponds to (e), and ultrasubharmonics of order $5/3$. All of these portraits are similar to each other, but it should be noted that for ultrasubharmonics of order m/n, when either m or n is even, there appear a pair of attractors and basins, while in cases where both m and n are odd, only one such attractor appears. This results from the symmetry of the system explained above.

Figure 18 shows an attractor-basin portrait corresponding to the point (j) on the kB chart of Fig. 7; see also [32]. There are two attractors of two-periodic groups; their two basins are separated by the ω-branches of a directly unstable fixed point of f_λ. Furthermore, each basin of a two-periodic point is subdivided by ω-branches of the inversely unstable fixed points. The tails of these four subdivisions of the two basins wind around the two-periodic groups respectively, becoming infinitely thin as they accumulate along the ω-branches of the directly unstable fixed point.

Figure 19 corresponds to the point (n) of Fig. 7. In this example, there exists only one attracting three-periodic group. The three images of this three-periodic point are considered as distinct attractors of the mapping f_λ^3, and the corresponding basins are distinguished. As is seen in the figure, a more complicated configuration appears, that is, each basin of an image of the three-periodic point is bounded by ω-branches of the directly unstable three-periodic points and the tails of these three basins behave in a complicated fashion, seen in the figure, becoming infinitely thin as they accumulate along the ω-branches of the three fixed points (one directly and two inversely unstable saddles). Moreover, tails of the basins are mixed in confusion with each other and accumulate on the ω-branches of these three saddles from both sides. It should be added that a closer inspection reveals that the basin boundaries of Fig. 19 have a fractal structure, but those of Fig. 18 do not. This means that invariant branches of the directly unstable saddles of Fig. 18 are heteroclinic but not homoclinic, while those of Fig. 19 are homoclinic.

Figures 20 and 21 correspond to (q) and (t), respectively. In addition to the completely stable fixed point, there exist two two-periodic groups in Fig. 20 and one three-periodic group in Fig. 21. Below the frame of Fig. 20 there exists a directly unstable fixed point as shown in the figure. Though not indicated in the figure, it has been confirmed that the invariant branches of this saddle have a homoclinic structure. In Fig. 20, the ω-branches delimiting basin boundaries of two-periodic points do not have homoclinic structure, but we have confirmed that these basins are caught in the homoclinic structure of the saddle below the frame, and hence tails of the basins show extremely complicated shape as we see in the figure. This case shows a second mechanism for generating a fractal basin boundary.

Taking the saddle below the frame of Fig. 20 into account, the Levinson–Massera relation (13) concerning the number of the fixed points holds and also equations (14) and (15) can be verified for the n-periodic points $(n = 2, 3)$ for all attractor-basin portraits given above.

Within each region in Fig. 7 identified by Roman numerals, the number of fixed points is constant, but the number changes across the boundaries of these regions. It is believed that all fixed points which can exist within the parameter range of Fig. 7 have been identified, although we cannot exclude the possibility that some fixed points existing in some small region at low damping may have been neglected. From Fig. 8 and the attractor-basin portraits of Figs 13–21 it is conjectured that the counts of fixed points in Table 1 are complete within the indicated regions. Recalling that there are no sources, we use $S(1)$ to denote the number of period one sinks; as before, $D(1)$ and $I(1)$ denote numbers of directly and inversely unstable saddle fixed points.

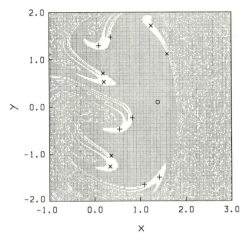

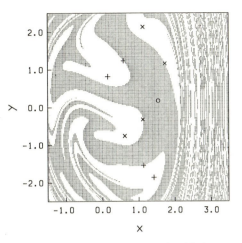

Fig. 15. Attractor-basin phase portrait with integration step size $2\pi/60$ and 201×201 grid of initial conditions for case (c) of Fig. 7; $k = 0.015$, $B = 0.45$. (Reproduced with the courtesy of the European Conference on Circuit Theory and Design [28].)

Fig. 16. Attractor-basin phase portrait with integration step size $2\pi/60$ and 201×201 grid of initial conditions for case (d) of Fig. 7; $k = 0.04$, $B = 0.90$. (Reproduced with the courtesy of the European Conference on Circuit Theory and Design [28].)

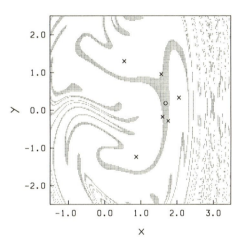

Fig. 17. Attractor-basin phase portrait with integration step size $2\pi/60$ and 201×201 grid of initial conditions for case (e) of Fig. 7; $k = 0.05$, $B = 1.40$. (Reproduced with the courtesy of the European Conference on Circuit Theory and Design [28].)

Table 1. Number of fixed points in the xy plane for various parameter regions

Region in kB chart	Number of fixed points	
I	$S(1) = 2$,	$D(1) = 1$
II–II′–III	$S(1)$ or $I(1) = 2$,	$D(1) = 1$
II′	$S(1) = 4$,	$D(1) = 3$
II ∩ III	$S(1) = 1$,	$S(1)$ or $I(1) = 2$, $D(1) = 2$
III–II	$S(1) = 2$,	$D(1) = 1$
IV	$S(1)$ or $I(1) = 2$,	$D(1) = 1$
Outside I ∪ II ∪ III ∪ IV	$S(1) = 1$	

171

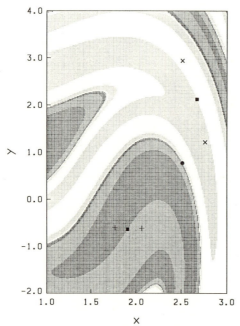

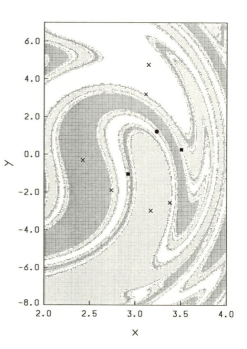

Fig. 18. Attractor-basin phase portrait with integration step size $2\pi/90$ and 201×301 grid of initial
conditions for case (j) of Fig. 7; $k = 0.20$, $B = 5.50$.
(Reproduced with the courtesy of the European Conference on Circuit Theory and Design [28].)

Fig. 19. Attractor-basin phase portrait with integration step size $2\pi/120$ and 201×301 grid of initial
conditions for case (n) of Fig. 7; $k = 0.20$, $B = 10.0$.
(Reproduced with the courtesy of the European Conference on Circuit Theory and Design [28].)

8. GLOBAL ATTRACTOR STRUCTURE, ATTRACTOR AND BASIN BIFURCATIONS

Let us now interpret the experimental facts described above in light of the theory,
particularly the global invariant manifold structures.

Figure 22 shows the chaotic attractor for case (o) in the kB plane, and some of the
associated invariant manifold structure. There are three unstable fixed points contained in
the attractor, one directly unstable point $^1D^1$ and two inversely unstable points $^1I^1$ and
$^2I^1$. The α- and ω-branches of the directly unstable saddle are shown in part, and a
number of transverse homoclinic intersections can be seen. As more of these branches are
prolonged, self-similar property will appear. Following Birkhoff, this implies the existence
of infinitely many periodic points near the homoclinic points.

Numerical evidence strongly suggests that the chaotic attractor is identical with the
closure of the α-branches or unstable manifolds of the directly unstable saddle $^1D^1$. This
phenomenon is not peculiar to the specific parameters of case (o), but occurs for typical
parameter values in the corresponding shaded region. The appearance of the attractor
seems to vary continuously as the parameter values are changed inside the shaded region
near the point (o).

Most homoclinic intersections in Fig. 22 are clearly transverse, but there are a few
instances of intersections which are very nearly tangent. It is expected that many such
tangencies or near tangencies may appear as more of the invariant manifolds are

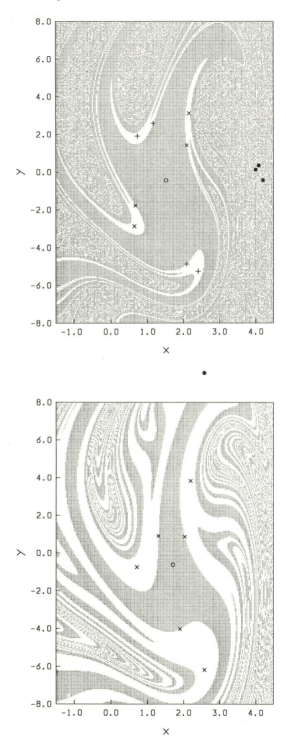

Fig. 20. Attractor-basin phase portrait with integration step size $2\pi/180$ and 241×321 grid of initial conditions for case (q) of Fig. 7; $k = 0.07$, $B = 16.1$. (Reproduced with the courtesy of the European Conference on Circuit Theory and Design [28].)

Fig. 21. Attractor-basin phase portrait with integration step size $2\pi/180$ and 241×321 grid of initial conditions for case (t) of Fig. 7; $k = 0.12$, $B = 18.8$. (Reproduced with the courtesy of the European Conference on Circuit Theory and Design [28].)

224   YOSHISUKE UEDA

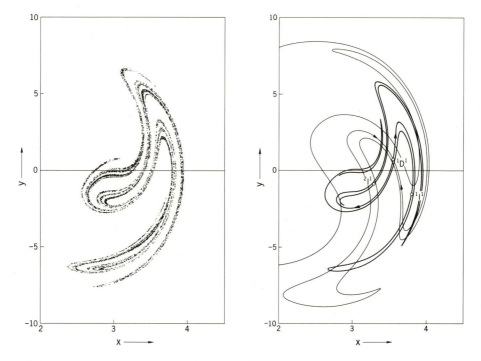

Fig. 22. Chaotic attractor for case (o) and the associated invariant manifold structure. (Reproduced with the courtesy of the Institute of Electrical Engineers of Japan [23].)

constructed. This suggests that the attractor structure may be structurally unstable in the sense of Andronov–Pontryagin.

Numerical experiments show that all of the infinitely many periodic motions in the attractor are unstable. Thus if any sinks exist, their basins of attractions are so small that the smallest amount of noise or error perturbs the system away.

The movement of images in the attractor under iteration is not reproducible, in the sense that nearly identical initial conditions lead eventually to different motions or waveforms. Furthermore, this situation occurs for motions starting from any part of the attractor; this property is referred to as sensitive dependence on initial conditions. On the other hand, any single long-term trajectory will, after transients die away, fill out an apparently identical structure, that is, the closure of the α-branches of $^1D^1$. Thus the numerical evidence strongly suggests that there is a single transitive attractor. A typical orbit returns infinitely often to a neighborhood of any point in the attractor, i.e. there is stability in the sense of Poisson.

Thus the observed motion may be thought of as visiting neighborhoods of various unstable periodic motions, which are infinite in number. The system continues to transit in apparently random manner among the infinitely many unstable periodic solutions. The transit may be influenced by small fluctuations or noise not included in the differential equations but present in the real system or simulation.

For this reason we have called this type of motion a randomly transitional motion [33]. The term chaos has since been widely accepted to describe this type of motion; this term strongly indicates the random aspect of the phenomena, although it may not adequately

174

convey the very coherent structure which is an equally important aspect of the motion.

Let us now turn to a description of the various routes to chaos observed as the system parameters are varied from outside into the shaded regions. The most commonly observed transition is by successive period doubling, beginning from the two symmetrically related sinks which exist everywhere inside region II. This first period doubling is illustrated in Fig. 8 from case (i) to case (j). The arc in Fig. 7 passing between points (i) and (j) shows the location where this first period doubling occurs. Notice that this period doubling bifurcation arc extends over and around to the right side of the shaded regions. A typical path in the kB plane which passes through this arc into the shaded regions leads to chaos via Feigenbaum cascade, generating two symmetrically related chaotic attractors which eventually merge with each other.

A different bifurcation is observed when entering the shaded (o) region from the right, that is, from the region typified by case (p). At this edge of the shaded (o) region, a global bifurcation occurs which causes the chaotic attractor to suddenly appear. The global bifurcation is a homoclinic tangency of the α- and ω-branches of the directly unstable fixed point (filled circle) in the basin boundary. This situation is illustrated in Figs 23–24. The parameter values $k = 0.1$, $B = 13.388$ are the values where the chaotic attractor gains or loses stability, and also the values where the directly unstable fixed point has a homoclinic tangency. This phenomenon was originally described in [34] and called a transition chain; similar phenomena have since been widely reported and are now frequently referred to as boundary crises [35] or blue sky catastrophes [36].

Another example of such a global bifurcation occurs along the right boundary of the shaded region containing case (l). Here the chaotic attractor gains or loses stability as it touches a directly unstable three-periodic point, whose α- and ω-branches develop a homoclinic tangency at the bifurcation threshold. This situation is illustrated in Fig. 25.

Note that in both this and the preceding case, the directly unstable point in the basin boundary has no homoclinic structure before the transition chain is established. This means that the chaotic attractor always has a regular basin boundary. In other systems, it can happen that a chaotic attractor exists in a basin with a fractal boundary, that is, homoclinic structure develops in the basin boundary prior to the transition chain or blue sky catastrophe. Further description of this phenomenon can be found in [37], where the term chaotic saddle catastrophe was used; see also [38]. The phenomenon of chaotic saddle catastrophe has not been observed for equation (16) in the region of the kB plane shown in Fig. 7.

Finally we describe the bifurcations passing from the region of case (n) to the right into the shaded region typified by case (o). Here the phenomena are somewhat more complicated. The chaotic attractor develops from the period three sink of case (n). First this period three sink experiences a period doubling cascade which leads to a three-piece chaotic attractor; then there is usually a sudden explosion in size from a three-piece to one large chaotic attractor. Similar phenomena were reported in [39], and have since become widely known as interior crises [35]; see also [13].

We note that in all cases where chaotic attractors are observed, they contain among the periodic points of lowest period either one inversely unstable point, or one directly unstable and two inversely unstable points. Recently Stewart has conjectured that this can be explained by applying the Levinson–Massera index equations to an appropriate subregion of the xy plane [40].

Sometimes the global structure of the phase portrait may have important consequences even though all attractors are regular periodic motions. For example, a local saddle-node or fold bifurcation causes the system to undergo a rapid transient to some other attractor; and it can happen that if two or more attractors are available, the one chosen may depend

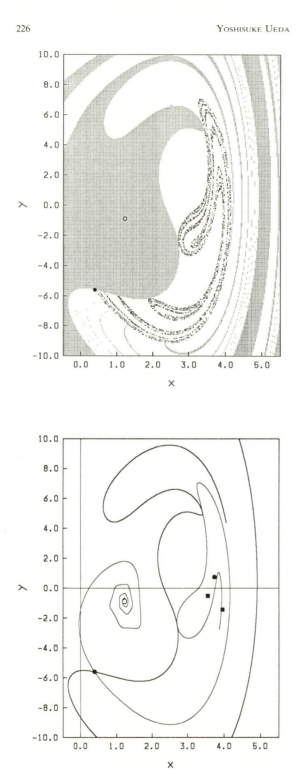

Fig. 23. Attractor-basin phase portrait with integration step size $2\pi/180$ and 241×321 grid of initial conditions for $k = 0.1$, $B = 13.388$.

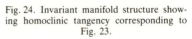

Fig. 24. Invariant manifold structure showing homoclinic tangency corresponding to Fig. 23.

176

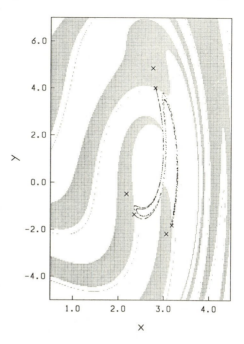

Fig. 25. Attractor-basin phase portrait with integration step size $2\pi/120$ and 201×301 grid of initial conditions for $k = 0.25$, $B = 8.812$.

very sensitively on how the bifurcation is realized in the simulation or real-world system. Such indeterminate bifurcations can be observed in the Duffing equation (16), for example by crossing the boundary of region I in the kB plane. Although the bifurcation event is local, the outcome is determined by global structure: the probabilities of settling on the various attractors available can be estimated from the structure of invariant manifolds; see [41].

9. CONCLUSION

By summarizing the author's previous reports [23–24, 28, 33–34, 39], regular and chaotic phenomena have been surveyed which occur in the system governed by the Duffing equation. The global attractor structure, and attractor and basin bifurcations have been discussed in relation to the geometric theory of differential equations. Chaotic attractors have been observed over a wide range of parameter values. Multiple coexisting attractors are also common; both regular and fractal basin boundaries are observed. The structure of chaotic attractors, basin boundaries and bifurcations have been understood in terms of unstable periodic motions and invariant manifolds.

High resolution portraits of chaotic attractors shown in Fig. 9(k) and (o_1) were lucky enough to earn the attention of a wide audience: D. Ruelle named the attractor of Fig. 9(o_1) *Japanese attractor* [5, 6], and Thompson and Steward referred to the attractor of Fig. 9(k) as *Ueda's chaotic attractor* [2]. Ueda's chaotic attractor shown in Fig. 26 and the Japanese attractor in Fig. 27 conclude the article. Both pictures are depicted taking the unit length of the x-axis to be five times that of the y-axis, and 100,000 steady state points are plotted.

177

Acknowledgement—I owe sincere thanks to the mathematician Dr Hugh Bruce Stewart of the Division of Applied Science, Brookhaven National Laboratory, for his generous assistance in preparing this article. Not only did he graciously agree to edit my manuscript, which was written in poor English, but he also provided me with much valuable advice in structuring the text as well as selecting the figures. I would like to point out that most of the items within the text that are not specifically described in the References [23–24, 28, 33–34, 39], are largely a result of his advice.

I would also like to thank Dr Stewart as well as Professor Ralph Abraham of the University of California Santa Cruz, especially for their insight into the significance of C. Pugh's closing lemma, which I had not fully grasped.

Fig. 26. Ueda's chaotic attractor.

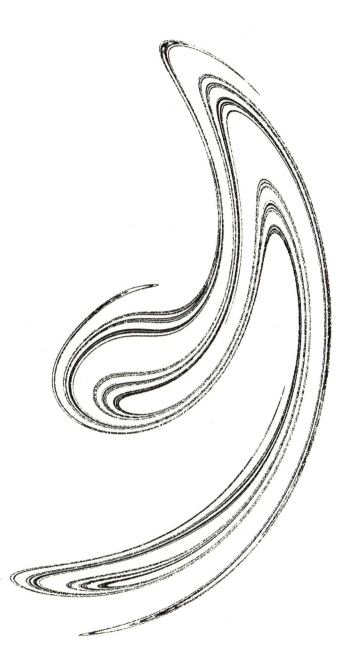

Fig. 27. Japanese attractor.

REFERENCES

1. G. Duffing, *Erzwungene Schwingungen bei veränderlicher Eigenfrequenz und ihre technische Bedeutung*. Vieweg, Braunschweig (1918).
2. J. M. T. Thompson and H. B. Stewart, *Nonlinear Dynamics and Chaos*. Wiley, New York (1986).
3. D. Ruelle and F. Takens, On the nature of turbulence, *Commun. Math. Phys.* **20**, 167–192; **23**, 343–344 (1971).
4. C. Hayashi, Y. Ueda, N. Akamatsu and H. Itakura, On the behavior of self-oscillatory systems with external forcing. *Trans. Inst. Commun. Engrs* **53A**, 150–158 (1970) (in Japanese); English Translation, *Electronics and Communications in Japan*, pp. 31–39. Scripta, Silver Spring, MD.
5. D. Ruelle, Les attracteurs étranges, *La Recherche* **11**, 132–144 (1980).
6. D. Ruelle, Strange attractors, *The Mathematical Intelligencer* **2**, 126–137 (1980).
7. H. Poincaré, *Les Méthodes Nouvelles de la Mécanique Céleste*, Vols 1–3. Gauthier-Villars, Paris (1899).
8. G. D. Birkhoff, *Collected Mathematical Papers*. American Mathematical Society, Providence, RI (1950); Republication, Dover Publications, New York (1968).
9. E. A. Coddington and N. Levinson, *Theory of Ordinary Differential Equations*. McGraw-Hill, New York (1955).
10. V. V. Nemytskii and V. V. Stepanov, *Qualitative Theory of Differential Equations*. Princeton University Press, Princeton, NJ (1960).
11. R. Abraham, *Foundations of Mechanics*. Benjamin, New York (1967).
12. R. Abraham and J. E. Marsden, *Foundations of Mechanics*. Benjamin/Cummings, Massachusetts (1978).
13. R. H. Abraham and C. D. Shaw, *Dynamics: The Geometry of Behavior, Part One, Periodic Behavior* (1982); *Part Two, Chaotic Behavior* (1983); *Part Three, Global Behavior* (1985); *Part Four, Bifurcation Behavior* (1988). Aerial Press, Santa Cruz, CA.
14. J. Guckenheimer and P. Holmes, *Nonlinear Oscillations, Dynamical Systems, and Bifurcations of Vector Fields*. Springer, New York (1983).
15. F. C. Moon, *Chaotic Vibrations, An Introduction for Applied Scientists and Engineers*. Wiley, New York (1987).
16. D. K. Arrowsmith and C. M. Place, *An Introduction to Dynamical Systems*. Cambridge University Press, Cambridge (1990).
17. N. Levinson, Transformation theory of non-linear differential equations of the second order, *Ann. Math.* **45**, 723–737 (1944); Correction **49**, 738 (1948).
18. N. Levinson, On the existence of periodic solutions for second order differential equations with a forcing term, *J. Math. Phys.* **22**, 41–48 (1943).
19. N. Levinson, On a non-linear differential equation of the second order, *J. Math. Phys.* **22**, 181–187 (1943).
20. C. C. Pugh, An improved closing lemma and a general density theorem, *Am. J. Math.* **89**, 1010–1021 (1967).
21. J. L. Massera, The number of subharmonic solutions of non-linear differential equations of the second order, *Ann. Math.* **50**, 118–126 (1949).
22. E. C. Zeeman, Bifurcation and catastrophe theory, in *Contemporary Mathematics*, Vol. 9, pp. 207–272. American Mathematical Society, Providence, RI (1982).
23. Y. Ueda, Random phenomena resulting from nonlinearity, *Trans. Inst. Electrical Engrs* **98A**, 167–173 (1978) (in Japanese); English Translation, *Int. J. Non-Linear Mech.* **20**, 481–491 (1985).
24. Y. Ueda, Steady motions exhibited by Duffing's equation: a picture book of regular and chaotic motions, in *New Approaches to Nonlinear Problems in Dynamics*, edited by P. J. Holmes, pp. 311–322. SIAM, Philadelphia (1980).
25. C. Hayashi, *Forced Oscillations in Non-Linear Systems*. Nippon Printing and Publishing Co., Osaka, Japan (1953).
26. C. Hayashi, *Nonlinear Oscillations in Physical Systems*. McGraw-Hill, New York (1964).
27. C. Hayashi, *Selected Papers on Nonlinear Oscillations*. Nippon Printing and Publishing Co., Osaka, Japan (1975).
28. Y. Ueda and S. Yoshida, Attractor-basin phase portraits of the forced Duffing's oscillator, *Proc. European Conf. Circuit Theory Design*, Paris, Vol. 1, pp. 281–286 (1987).
29. C. Hayashi, Y. Ueda and H. Kawakami, Transformation theory as applied to the solutions of non-linear differential equations of the second order, *Int. J. Non-Linear Mech.* **4**, 235–255 (1969).
30. C. Hayashi, Y. Ueda and H. Kawakami, Periodic solutions of Duffing's equation with reference to doubly asymptotic solutions, *Proc. 5th Int. Conf. Nonlinear Oscillations*, Kiev, Vol. 2, pp. 507–521 (1970).
31. C. Hayashi and Y. Ueda, Behavior of solutions for certain types of nonlinear differential equations of the second order, *Proc. 5th Int. Conf. Nonlinear Oscillations*, Poznań, Vol. 14, pp. 341–351 (1973).
32. C. Hayashi, Y. Ueda and H. Kawakami, Solution of Duffing's equation using mapping concepts, *Proc. 4th Int. Conf. Nonlinear Oscillations*, Prague, pp. 25–40 (1968).
33. Y. Ueda, N. Akamatsu and C. Hayashi, Computer simulation of nonlinear ordinary differential equations and non-periodic oscillations, *Trans. Inst. Commun. Engrs* **56A**, 218–225 (1973) (in Japanese); English Translation, *Electronics and Communications in Japan*, pp. 27–34. Scripta, Silver Spring, MD.
34. Y. Ueda, Randomly transitional phenomena in the system governed by Duffing's equation, *J. Statistical Phys.* **20**, 181–196 (1979).

35. C. Grebogi, E. Ott and J. A. Yorke, Crises, sudden changes in chaotic attractors, and transient chaos, *Physica* **7D**, 181–200 (1983).
36. R. H. Abraham, Chaostrophes, intermittency, and noise, in *Chaos, Fractals, and Dynamics*, edited by P. Fischer and W. R. Smith, pp. 3–22. Marcel Dekker, New York (1985).
37. H. B. Stewart, A chaotic saddle catastrophe in forced oscillators, in *Dynamical Systems Approaches to Nonlinear Problems in Systems and Circuits*, edited by F. M. A. Salam and M. L. Levi, pp. 138–149. SIAM, Philadelphia (1988).
38. C. Grebogi, E. Ott and J. A. Yorke, Basin boundary metamorphoses: changes in accessible boundary orbits, *Physica* **24D**, 243–262 (1987).
39. Y. Ueda, Explosion of strange attractors exhibited by Duffing's equation, *Ann. NY Acad. Sci.* **357**, 422–434 (1980).
40. H. B. Stewart, Application of fixed point theory to chaotic attractors of forced oscillators, Research Report, NIFS-62, National Institute for Fusion Science, Nagoya, Japan (1990); *Japan J. Ind. Appl. Math.*, to appear.
41. H. B. Stewart and Y. Ueda, Catastrophes with indeterminate outcome, *Proc. R. Soc.* **A432**, 113–123 (1991).

Selection 7

STRANGE ATTRACTORS AND
THE ORIGIN OF CHAOS[*]

Yoshisuke Ueda

Department of Electrical Engineering
Kyoto University, Kyoto 606, Japan

1. PROLOGUE

I am greatly honored to have been given this wonderful theme, "Strange Attractors and the Origin of Chaos" for my presentation today. First I would like to take this opportunity to offer a special thanks to each one of the people who planned and made this symposium possible.

At present, people say that the data I was collecting with my analog computer on the 27th of November, 1961, is the oldest example of chaos discovered in a second-order non-autonomous periodic system. Around the same time, it was Lorenz who made the discovery of chaos in a third-order autonomous system.

At that time I was simply frustrated with this seemingly mysterious phenomenon which I accidentally came upon during my experiments. For my part, it was nothing as glorious as an act of discovery — all I did for a long period of time was to keep on pursuing my stubborn desire to understand this unsettling phenomenon. "What are the possible steady states of a nonlinear system?" —— this has always been my question. And my paradigm has always been the phenomena themselves, not papers with their abstractions, but something we can actually observe or quantify.

In this report, I would like to reflect upon the circumstances of my research and the general conditions of Japanese science around 1978 before the study of

[*]This article was first presented at the international symposium entitled "The Impact of Chaos on Science and Society," where I was invited as a guest speaker. The symposium was organized by the United Nations University and the University of Tokyo, and held in Tokyo between 15-17 April, 1991. I am greatly indebted to the United Nations University for giving me their permission to publish this article in this volume and in "Nonlinear Science Today," Vol. 2, No. 2 (1992) prior to the publication of the proceedings of the symposium.

chaos began. As I prepared this talk, I kept asking myself what propelled me to pursue my question so relentlessly, but I must confess that I don't know the answer. I have not, in my wildest dreams, imagined that I would be given an opportunity to speak on this very subject. It was so unexpected, my mind was reduced to total chaos!

As an academic term, I do not find the word "chaos" very appropriate. But what shall we call it then? My proposal has been "randomly transitional phenomena"; I will explain this below.

The characteristics of chaos in a physical system can be summarized as follows:

Random phenomena that occur in deterministic systems.

Random phenomena whose short-term behavior is predictable.

Random phenomena whose long-term behavior is unpredictable.

Although the phenomena are irregular and unpredictable, chaos
does have a definite structure.

The original meaning of chaos, I feel, is "total disorder and ultimate unpredictability." But as scientific terminology, the word "chaos" seems to overemphasize the unpredictability alone. This symposium provides an opportunity to clear this misunderstanding and to inform the non-specialist of the correct meaning of the word.

Even so I have to use the word "chaos" here, instead of "randomly transitional phenomena." It is a concise expression which has already filtered into people's minds, and therefore I have decided it is rather pointless to resist it.

2. THE OLDEST CHAOS IN A NON-AUTONOMOUS SYSTEM — A SHATTERED EGG

The 27th of November, 1961 became a memorable day for me, although I did not have any joyous realization that something wonderful had happened, or any vivid memory of the events of that particular day. As I remember, I had just finished writing the narrative to accompany the data I was going to publish at the Special Committee on Nonlinear Theory of the Institute of Electrical Communication Engineers, to be held on December 16th, and was carrying out some analog computer experiments with the help of Susumu Hiraoka, who was two years my junior, in order to test the applicability of the approximate computation I quoted in my paper. Had I not had the date on the printout of that old analog computer, which was destined for a wastebasket, I would never have been able to recall the date (Fig. 1).

At that time, I was a third-year graduate student at Kyoto University, working on the phenomenon of frequency entrainment under the guidance of Professor Chihiro Hayashi. When a circuit (oscillator) which would, if left alone, keep on generating an electrical (self-sustained) oscillation with a certain frequency and amplitude, is driven externally with signals whose frequency is different from that of the self-oscillator, its self-oscillating frequency is drawn to and synchronized with that of the driving frequency. This phenomenon is called frequency entrainment. There are exceptions, of course — depending on the value of the driving frequency and amplitude, entrainment sometimes does not occur. Instead, an aperiodic "beat oscillation" with drifting frequencies would appear.

I was receiving direct guidance from Professor Hiroshi Shibayama of Osaka Technological University, who was visiting our laboratory as a guest scholar several times a week. Warm and gentle, he let me do whatever I wanted to do, while introducing me to the basics of the research. Even now I look up to him as my senior friend and receive all kinds of advice.

The main purpose of my computer experiment was to simulate the non-autonomous nonlinear differential equation describing frequency entrainment, and to examine the range of the frequency and amplitude of the driving signals which cause synchronization, as well as the amplitude and the phase of its oscillation. Allow me to go into a little technical detail here. The approximate computation was done by rewriting the non-autonomous equation into an autonomous equation using the averaging method (approximation enters the process at this stage, with the result that chaos is suppressed). The aim was to approximate the steady state of the original system with an equilibrium point or limit cycle of the autonomous equation. In this method, the stable equilibrium point corresponds to synchronized frequency entrainment, and a stable limit cycle corresponds to asynchronous drift conditions. Actually there are two kinds of asynchronous oscillations —— quasi-periodic oscillation (represented by a limit cycle in the averaged equation) and chaotic oscillation (represented by strange or chaotic attractors): but the common sense of the day failed to recognize the chaotic oscillation. In those days only the equilibrium point and limit cycle were known to exist as steady states of a (second order) autonomous system, so it was understandable for everyone to have possessed the preconceived notion that asynchronous condition meant quasi-periodicity. On that day, the 27th of November, when I changed the parameter (frequency of the driving input), and the condition shifted from frequency entrainment to asynchronization, the oscillation phenomena portrayed by my analog computer was chaotic indeed. It was nothing like the smooth oval closed curves in Fig. 1,

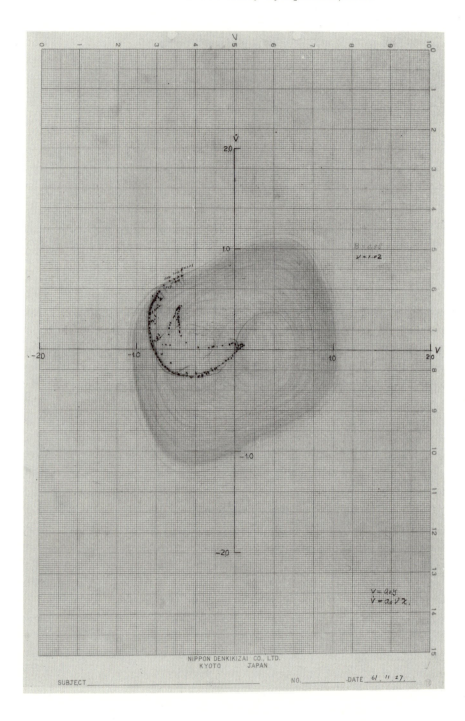

but was more like a broken egg with jagged edges. My first concern was that my analog computer had gone bad. But I soon realized that that was not the case. It did not take long for me to recognize the mystery of it all –– the fact that during the asynchronous phase, the shattered egg appeared more frequently than the smooth closed curves, and that the order of the dots which drew the shattered egg was totally irregular and seemingly inexplicable. As I watched my professor preparing the report without a mention of this shattered egg phenomenon, but rather replacing it with the smooth closed curves of the quasi-periodic oscillation, I was quite impressed by his technique of report writing. But at the same time, I realized that one needs to be very careful in reading reports of this sort [1].

I am getting a little sidetracked here, but the analog computer had been developed and created as his research project by Minoru Abe who was three years my senior. My deep respect goes to him who so laboriously and meticulously handbuilt this practical computer with vacuum tubes. Figure 2 is an example of the waveform data made by the computer. It reminds me of the long hours patiently sitting in front of the analog computer, and of my wonder at its accuracy — a testament to its maker's unique skill. To obtain a sheet of data as shown in Fig. 2, it took the computer about 60 to 100 minutes. Most of them have been discarded, but we had accumulated at least 1000 sheets of data during my five years in graduate school. I would like to mention that I had not contributed very much in creating this computer beyond helping him build several operational amplifiers and learning to repair the chopper amplifier and the recorder. As I look back, I feel that after those long exhausting vigils in front of the analog computer, staring at its output, chaos had become a totally natural, everyday phenomenon in my mind. People call chaos a new phenomenon, but it has always been around. There's nothing new about it — only people did not notice it.

Fig. 1 Output of an analog simulation of the equation

$$\ddot{v} - \mu(1 - \gamma v^2)\dot{v} + v^3 = B \cos \nu t \qquad (1)$$

with $\mu = 0.2$, $\gamma = 8$ and $B = 0.35$ obtained on 27 November 1961 is shown. A continuous orbit is drawn lightly on the $v\dot{v}$ plane and points in the Poincaré section at phase zero are given by heavy dots; five dots near the top are fixed points for a sequence of values at $\nu = 1.01$, 1.012, 1.014, 1.016 and 1.018, the remaining points are on the chaotic attractor at $\nu = 1.02$.

(Courtesy of Dr. H. B. Stewart, Brookhaven National Laboratory)

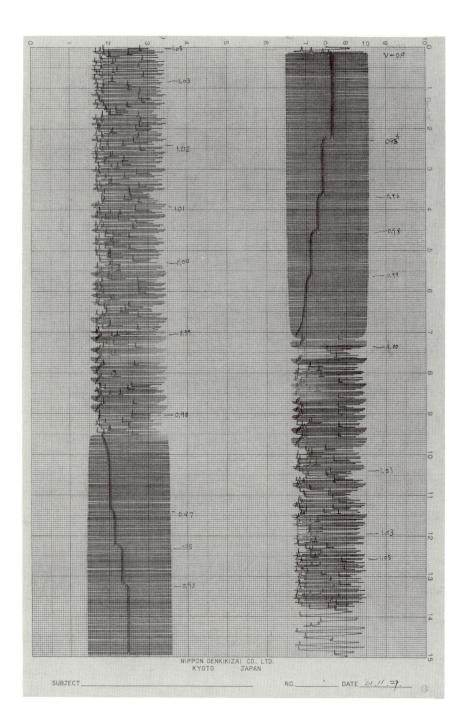

When I think of those long hours in front of the analog computer, I also think of my classmate, Toshiaki Murakami. He had completed his master's degree and came to work in April of 1961. During our graduate school days, we used to compete for the use of the analog computer. He and his instructor, Yoshikazu Nishikawa, four years ahead of us, used to keep long vigils, too, painstakingly examining point by point the domain of the initial conditions of the Duffing equation. Their main results appear in Fig. 10.6 in [2], and Fig. 4 of [3] in the bibliography, reproduced here as Fig. 3. In Figure 3 of [3], reproduced below as Fig. 4, we can clearly see the curve (alpha-branch) which swirls down from the saddle point 3 on the basin boundary to point 1. At the time, the pertinence of the figure was controversial among the specialists. Dr. Hayashi himself voiced his doubts and did not include it in his book published by McGraw-Hill in 1964. I will discuss this point later in section 5.

In the research data [1] published in December of 1961, the saddle-node point corresponding to the boundary of the frequency entrainment domain was plotted on the limit cycle, but a careful examination proved that it was actually not on the limit cycle. This error was corrected immediately after the publication.

The fact that these data survived was by itself a miracle. It symbolizes my packrat tendency — a typical trait for those who grew up during the war. Naturally I cannot verify the dates for those which were not dated at the time, but the following account can be corroborated from other data such as those of the nonlinear research group, to within a few months. I will touch upon the original data at the time of the experiments later in section 8.

Thus ended 1961. In 1962, the nonlinear oscillation group of Hayashi students was studying frequency entrainment phenomenon of the self-oscillatory system with driving periodic signals (forced self-oscillatory system) and the oscillatory phenomenon of a system which holds a steady equilibrium if no ex-

Fig. 2 Waveforms of an analog simulation of Eq. (1) with varying ν are shown for $B = 0.3$. A pulse on the waveform which is generated by an auxiliary circuit independent of the main analog computation circuit for Eq. (1) shows a time mark at every period of the external periodic forcing; when this pulse sequence forms a straight line (after transient), we can infer that periodic motion appears with the system being entrained by the external periodic signal. These data progress from entrained state to asynchronized state and vice versa exhibiting a narrow hysteresis zone.

(Courtesy of Dr. H. B. Stewart, Brookhaven National Laboratory)

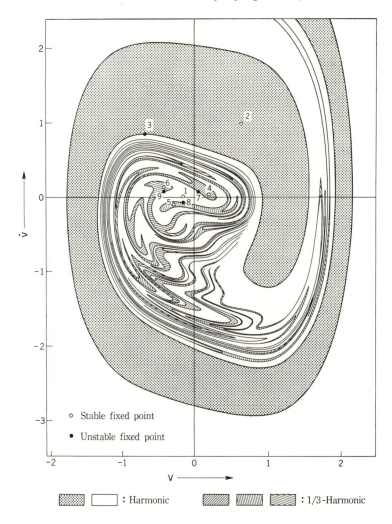

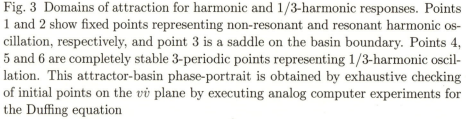

Fig. 3 Domains of attraction for harmonic and 1/3-harmonic responses. Points 1 and 2 show fixed points representing non-resonant and resonant harmonic oscillation, respectively, and point 3 is a saddle on the basin boundary. Points 4, 5 and 6 are completely stable 3-periodic points representing 1/3-harmonic oscillation. This attractor-basin phase-portrait is obtained by exhaustive checking of initial points on the $v\dot{v}$ plane by executing analog computer experiments for the Duffing equation

$$\ddot{v} + k\dot{v} + v^3 = B\cos t \tag{2}$$

with $k = 0.1$, $B = 0.15$.

(Reproduced with the courtesy of McGraw-Hill Book Company
and Nippon Printing and Publishing Company [2, 3, 17])

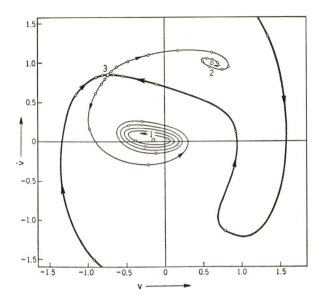

Fig. 4 The loci of some image points on the invariant curves of the directly unstable fixed point 3, for the same equation and parameter values as Fig. 3. (Reproduced with the courtesy of Nippon Printing and Publishing Company [3, 17])

ternal periodic signals are added (forced oscillatory system). The former research centered around Prof. Shibayama, focusing mainly on the analyses of Van der Pol equations with added forced terms, or of a mixture of Van der Pol and Duffing equations, while the latter, the study on the Duffing equation, centered around Prof. Nishikawa.

3. ENCOUNTER WITH THE JAPANESE ATTRACTOR

One day in the fall or winter of 1962, Prof. Chihiro Hayashi was waiting for me in the laboratory room. At that time he was working on the manuscript for his forthcoming book [2] which was eventually published in 1964 by McGraw-Hill. The laboratory room was probably more convenient for his work: he left his office vacant and occupied the laboratory room. "I want you to do this computation in a hurry," he told me. It was to solve third order simultaneous equations with four unknowns and draw the amplitude characteristic curve of the periodic solution of the Duffing equation, taking into consideration the components up to the third harmonic component. Let me explain the circum-

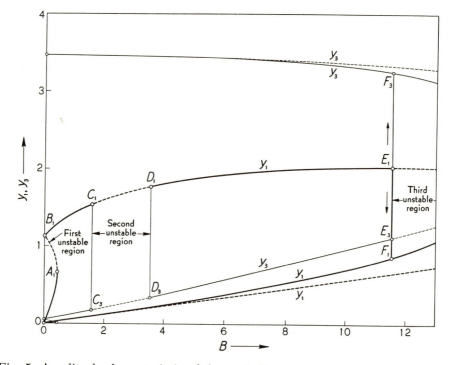

Fig. 5 Amplitude characteristic of the periodic solution

$$v = y_1 \cos t + y_3 \cos 3t \qquad (3)$$

of the Eq. (2) for $k = 0$, obtained by using harmonic balance method.
(Reproduced with the courtesy of Nippon Printing
and Publishing Company [4])

stances of Prof. Hayashi's request. When he sent chapter 6 of his book to McGraw-Hill, it was returned with an objection from a reviewer. The contents of chapter 6 were exactly the same as chapter 3 of his earlier book (1953) [4]. The reviewer, as I recall, was Prof. Higgins of the University of Wisconsin. In the book, Prof. Hayashi showed the amplitude characteristic curve for the periodic solution represented by cosine components alone, of the Duffing equation with zero as a damping term, (Fig. 5), but not the curve with non-zero damping coefficient. There was a reason for this. When the damping coefficient is not zero, sine components become necessary in the periodic solution, in addition to the cosine components, thus doubling our workload. It was particularly difficult on the hand-cranked calculator. I was somehow trusted by Prof. Hayashi as someone having fast and accurate computation and careful drawing skills. So on that occasion, I felt quite proud of myself and accepted the challenge,

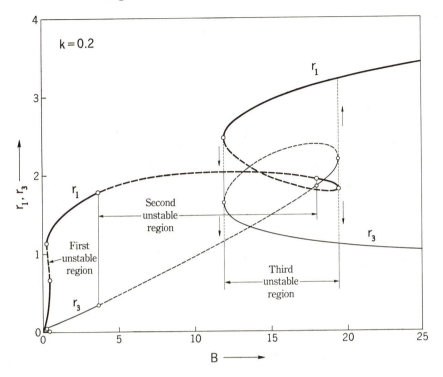

Fig. 6 Amplitude characteristic of the periodic solution

$$v = x_1 \sin t + y_1 \cos t + x_3 \sin 3t + y_3 \cos 3t \qquad (4)$$

with $r_1^2 = x_1^2 + y_1^2$, $r_3^2 = x_3^2 + y_3^2$, of the Eq. (2) for $k = 0.2$.
(Reproduced with the courtesy of McGraw-Hill Book Company [2])

knowing it was a difficult job. Even if I did not want to do it, I would not have been able to refuse him anyway. I was a little amused to find this most exalted professor high up in the clouds a bit flustered, especially with the objection written in English no less!

As it was impossible to solve the third order simultaneous equations with four unknowns by sheer brute force, I eliminated two variables from the four expressions to lead to high order simultaneous equations with two unknowns, gave the amplitude of the periodic solution ahead of time, with a bit of a "manipulation", returned to the computation of the amplitude of the external force, and finally, by hand calculations, solved approximately fifty cases of third order equations with the Newton-Raphson method, and drew the curves shown in Fig. 6 (Fig. 6.1 in the reference [2]). I fondly remember this crash project completed within a few weeks. The damping coefficient was set at 0.2.

195

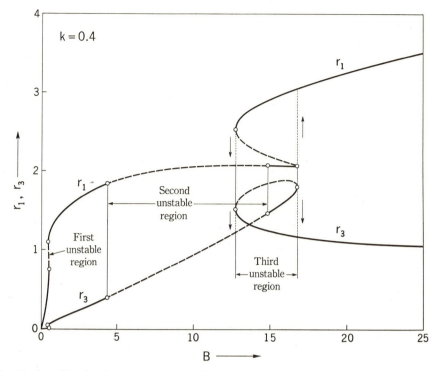

Fig. 7 Amplitude characteristic of the periodic solution

$$v = x_1 \sin t + y_1 \cos t + x_3 \sin 3t + y_3 \cos 3t \qquad (5)$$

with $r_1^2 = x_1^2 + y_1^2$, $r_3^2 = x_3^2 + y_3^2$, of the Eq. (2) for $k = 0.4$.
(Reproduced with the courtesy of Nippon Printing
and Publishing Company [6, 17])

The amplitude characteristic curves drawn with only two frequency components contained in the periodic solution, are nothing but an approximation. The standard procedure, therefore, was to verify the results with the analog computer. I struggled again for several days in a row, in front of the computer. When the amplitude of the external force increased, the high frequency components in the periodic solution also increased, accelerating the response time, thus making it extremely difficult for the servo multiplier — that represented the nonlinear term — to follow up. Consequently, one had to extend the computer's time scale so as to slow down the response time. I repeated the experiment with one cycle of the external force 2π corresponding to $31.4\cdots$ seconds. During the process, I encountered enough chaotic oscillations (the source of the Japanese attractor) to make me sick. But Prof. Hayashi told me, "Oh, it's probably taking time to settle down to the subharmonic oscillations.

196

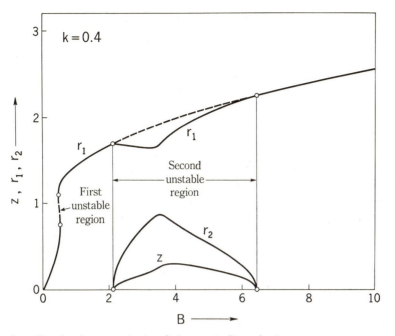

Fig. 8 Amplitude characteristic of the periodic solution

$$v = z + x_1 \sin t + y_1 \cos t + x_2 \sin 2t + y_2 \cos 2t \qquad (6)$$

with $r_1^2 = x_1^2 + y_1^2$, $r_2^2 = x_2^2 + y_2^2$, of the Eq. (2) for $k = 0.4$.
(Reproduced with the courtesy of Nippon Printing
and Publishing Company [6, 17])

Even in an actual series resonance circuit, such a transient state lingers for a long time." When I look back, though, I seemed to have sensed at that time that chaos was not a phenomenon unique to forced self-oscillatory systems in which quasi-periodic oscillation appeared.

Only a fragmental record of the original data of my work with the analog computer and manual calculation remains now. I may have been worried at that time that, had Prof. Hayashi seen this data, he would have told me to repeat the analog computer experiments until the transient state settled to a more acceptable result. Sensing that no matter how long I continued the simulation, I would never be able to come up with the data he wanted, short of making up some false data, I must have suggested the larger damping coefficient, 0.4, and done away with the problematical data, or burnt them in 1978 when we moved. These were the circumstances of obtaining the amplitude characteristic curve for my doctoral dissertation (Figs. 7 and 8). By the summer of 1963, the amplitude characteristic curves for the unsymmetrical periodic solution had already been obtained as well, but since they did not appear in the McGraw-Hill book, the manuscript editing must have been completed by that time. The Figures 7 and 8 were included in reference [5], but the description of chaos does not appear anywhere. Based on this data, Prof. Hayashi wrote his paper on higher-harmonic oscillation, which he reported during the International Congress at Marseilles [6]. I finished my doctorate in the spring of 1964, or more accurately, completed the units requirement, left school, and was hired as a research assistant in the Dept. of Electrical Engineering. During that year, Prof. Hayashi's book, **Nonlinear Oscillations in Physical Systems** came out [2]. Students are often critical of their teachers. I was thinking "the book may be nothing but a databook for harmonic balance methods." Now the table is turned. As I write this, I chide myself who has not yet published a book.

4. THE HAYASHI LABORATORY AT THE TIME OF THE "McGraw-Hill Book"

The above was the condition of the Hayashi Laboratory up until the publication of Prof. Hayashi's "McGraw-Hill Book." The laboratory was overflowing with chaotic data produced by the analog simulations, and yet they were overlooked as a quasi-periodic oscillation or transient state. I would like to touch upon the goings-on in the laboratory around that time. Prof. Shibayama, having finished his dissertation, had stopped coming around. Prof. Nishikawa had left nonlinear oscillations and changed his field to control engineering, a few years earlier. Prof. Abe was focusing his attention mainly on the application of analog techniques to control systems, and was concerned with nonlinear oscillations only indirectly. This made me the senior researcher on nonlinear oscillation in the laboratory.

Even after the publication of his book, Prof. Hayashi used to ask me to do all kinds of work for him. He was especially strict with me, always demanding "Don't give this work to anyone, be sure to do it yourself!" Even if he did not say that, there were several other senior researchers in the laboratory who used most of the students' help, leaving no one to help me.

Let me mention a group that started from then on. "From then on" because it continues even now. It includes Nobuo Sannomiya, three years my junior, and Masami Kuramitsu, four years my junior. Since 1963, we three took turns at holding seminars a couple of hours every week, reading Smirnov's **A Course of Higher Mathematics** (the original was in Russian). Starting in 1963, it took five years to finish studying the Japanese version from Vols. 5 through 12. It was truly helpful. Without the background thus developed by this seminar,

I, who had been trained with the engineering school curriculum would never have been able to understand papers such as Birkhoff's. Prof. Hayashi did not seem to like the idea of our round-robin seminar, and told us often to work on calculations if we had time to read books. But we kept ignoring his admonitions. He also had an extreme suspicion of and dislike for the digital computer KDC-I which finally became available around that time. But I used it to compare results such as the periodic solution of the Duffing equation with the analog data. The KDC-I was a machine built with transistors, and took about 60 seconds to integrate the Duffing equation along the time axis from t=0 to 2π, using the Runge-Kutta-Gill method that set the size of integration step at $2\pi/60$. It wouldn't even make a toy today, but at that time I was deeply impressed by the fact that such a calculation — impossible to do by hand — was finally possible.

5. FROM THE HARMONIC BALANCE METHOD TO THE MAPPING METHOD

Dr. Hayashi's "McGraw-Hill Book" was highly regarded, and he was invited to be a guest professor at Columbia University and the Massachusetts Institute of Technology from the fall of 1965 through the summer of 1966. I was his trusted student, and was even given the honor of handling his paychecks during his absence. As he was always present in the laboratory before and after his stay in the United States, I naturally cherished a year's freedom from his watchful eyes. That fall, during Prof. Hayashi's absence, Hiroshi Kawakami from Tokushima (five years my junior) was in the second year of his masters program, studying analog techniques and their applications to control systems. But he came to me for help in studying nonlinear oscillation which he apparently found interesting. It was impossible to change his research project in his second year of graduate school especially in such a small laboratory, but I drew him into the calculation and analog simulation of the amplitude characteristics of nonlinear oscillation with two saturable cores. Soon after, Prof. Hayashi returned. By then Kawakami was already in his doctorate program and his determination to pursue nonlinear oscillation research was already established. He was trying to select a thesis for his research. In our laboratory, as I have already mentioned, mapping or the stroboscopic method had been used for analog computer experiments, and technical terms such as completely stable fixed point were familiar to us. The knowledge came mostly from W. S. Loud's Papers [7]. Prof. Hayashi was friendly with Prof. Loud, regularly corresponding with him. Around that time we got a copy of N. Levinson's paper [8], although

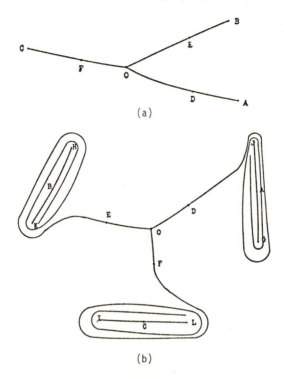

Fig. 9 Examples of maximum finite invariant domains.
   (a)   O: completely stable fixed point
         A, B, C: completely stable 3-periodic points
         D, E, F: directly unstable 3-periodic points
   (b)   O, D, E, F: the same as in (a)
         A, B, C: inversely unstable 3-periodic points
         G, H, I, J, K, L: completely stable 6-periodic points
(Reproduced with the courtesy of Annals of Mathematics [8])

I am not sure when and how it fell into our hands. Probably Kawakami, being a diligent student, copied it from somewhere. When I was studying the paper, I came upon the Figs. 2 and 3 on p. 735. The moment I understood the meaning of these figures, I thought "This is it!" (Fig. 9). It solved a long-standing mystery for me. Figure 4 was clearly an error, and the key to the correct answer was in these figures. I thought it through for several days, and figured it out also in numerical terms. This was my first encounter with the concept of heteroclinic point. Of course I did not know the term then. After several days, I was confident. I drew a rough sketch and presented it to Prof. Hayashi (Fig. 10). He did not agree with me immediately, but eventually did, and we decided to investigate all the cases in which subharmonic oscillation of order

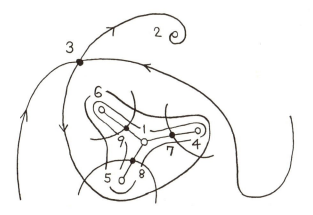

Fig. 10 Corrected schematic diagram of the invariant
curves corresponding to Fig. 4.

1/3 appeared. I recommended the subharmonic oscillation of order 7/3 which I had seen during my earlier analog simulation of higher harmonic oscillation. It used to appear where the external force was large and things got complicated, and I used to feel relieved. We immediately asked Kawakami to draw the invariant curves of the directly unstable fixed point in this case. At the same time we asked Prof. Abe to create an automatic mapping device. Figure 7(b) in the reference [9] shows our results. Figures 11, 12 and 13 were completed right after the abstracts for the 4th Conference on Nonlinear Oscillations held at Prague in 1967 had been mailed, and therefore became its appendix. Since these figures have never been published either in the proceedings of the Conference or in Kawakami's doctorate dissertation, except in the NLP research data [10], I would like to include them here. Please refer to references [10] and [11].

The first results we obtained from Prof. Abe's automatic mapping device are shown in Fig. 6 in reference [12] (reproduced here as Fig. 14), in which the error of Fig. 4 has been corrected. The device itself is summarized in the appendix of reference [13].

In modern terms, the automatic mapping device is a device which uses the analog computer to create Poincaré maps. In other words, the device plots on the recording paper a sampling of the analog signal at the same instant during each cycle of the external force. Although Kawakami and myself helped Prof. Abe build the trial device, it would have been extremely difficult without Prof. Abe's remarkable expertise. Thanks to the device, drawing invariant curves, or outstructures in more modern terms, became much easier. It goes without saying that as a result, the application of discrete dynamical systems theory to nonlinear oscillation was speeded up considerably. Through this experience I

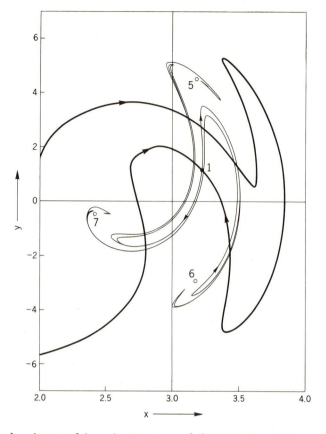

Fig. 11 Fixed points and invariant curves of the mapping T for equation

$$\dot{x} = y, \qquad \dot{y} = -ky - x^3 + B\cos t \qquad (7)$$

with $k = 0.2$ and $B = 10.0$.

(Unpublished supplements to Ref. [12])

learned the importance of examining the original texts as well as making the tools and devices on our own, which are necessary in achieving one's research goals.

In the spring of 1966, Norio Akamatsu (seven years my junior) came into the master's program. Another Tokushima native, he chose nonlinear oscillation as the subject of his research from the very start and came directly under my guidance. I asked him to begin with the application of the mapping method to the research in synchronization phenomenon, for which the harmonic balance method had been the standard method in the past. A part of his master's thesis is described in the latter half of reference [9]. These two students from

202

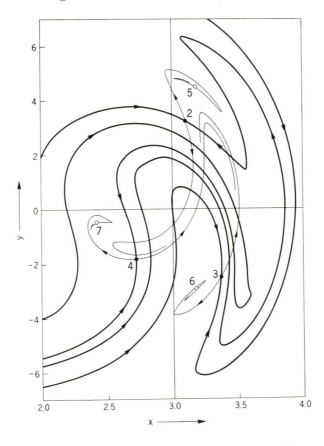

Fig. 12 Fixed points and invariant curves of the mapping T^3 for Eq. (7) with $k = 0.2$ and $B = 10.0$.

(Unpublished supplements to Ref. [12])

Tokushima were both extremely sharp, and contributed greatly to nonlinear oscillation research. I owe them a great deal. I did not force them to do anything under Prof. Hayashi's authority, but instead built a cooperative atmosphere so as to let their unique abilities grow to the fullest. I pride myself for helping to build the golden era of the Chihiro Hayashi Laboratory during the latter half of the 1960s. In fact, I tried to point out to them the particulars of bibliographic research and interpretations, while gathering data with them, examining their parameter settings, and at the same time acted as their protecting wall, or more accurately, I completed or did over their results which did not meet Prof. Hayashi's approval. Akamatsu was the last graduate student to complete his doctorate program in the Hayashi Laboratory. Kawakami left us in 1970, and Akamatsu left in 1971, leaving Masami Kuramitsu, Kenji

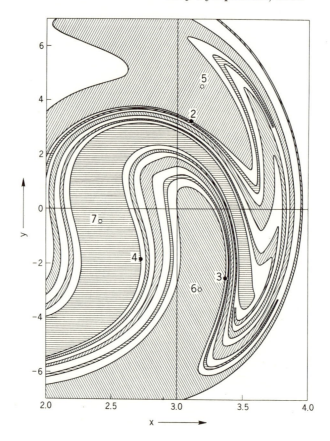

Fig. 13 Domains of attraction of the completely stable fixed points of the mapping T^3 for Eq. (7) with $k = 0.2$, $B = 10.0$.

(Unpublished supplements to Ref. [12])

Ohshima and myself, who were studying nonlinear oscillations in a postdoctorate capacity. Because of our research themes, we three maintained a parallel relationship. Kuramitsu was working on a nonlinear system with many degrees of freedom, while Ohshima was undertaking experimental research on actual electrical circuits.

6. THE TRUE VALUE OF AN ADVISOR: A SCION OF CHAOS

I cannot forget to mention my mentor, Dr. Michiyoshi Kuwahara, to whom my work is deeply indebted. Dr. Kuwahara was my senior fellow in the Hayashi

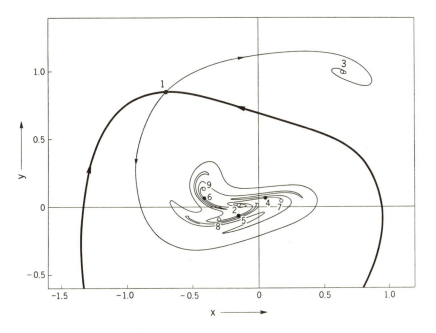

Fig. 14 Computed fixed points and invariant curves corresponding to Figs. 3, 4 and 10. Note: Numbering of the points is different, but the equation and the parameter values are the same.
(Reproduced with the courtesy of Nippon Printing
and Publishing Company [17])

Laboratory, and was the very first graduate student who completed his program in our Lab. He understood my position well, and had paid me a visit once a month for the past thirty some years, always ready to give me good advice and suggestions. To this day, I still listen to his wisdom. He does not mince his words —— a frankness I always like and admire. Sometimes his opinion struck close to home and was quite painful to my ear. But when I received a lashing from Prof. Hayashi, Dr. Kuwahara used to listen to me patiently and comfort me. I remember those occasions very vividly indeed.

I received much valuable advice from Dr. Kuwahara. Among his innumerable suggestions, I am most thankful for his insistence concerning the art of paper-writing. "Write papers and send them to appropriate academic journals." he advised. "Oral reports (nonlinear research group data, etc.) do not help you. Send them off yourself, so that it will be credited to you by the referees." I did not receive this kind of advice from Prof. Hayashi. Most of the data in the nonlinear research group reports were published by Prof. Hayashi at the international meetings, so I had to be careful not to duplicate his material.

Around the time when the university was in turmoil with student protest, I sent a paper of my own for the first time, to the **Journal of the Institute of Electronics and Communication Engineers**, which had referees. The democratic atmosphere that prevailed on campus because of the protest might have prompted me to take this action. The paper passed the review smoothly and was accepted. Figure 6 in the paper [14], demonstrates quasi-periodic oscillation and chaos in the Rayleigh-Duffing mixed type equation (Fig. 15). Though it hadn't taken a clear shape yet, the concept of chaos was already established in my mind by that time.

This fact can be supported by the following circumstance. A seminar called "Ordinary differential equation and nonlinear dynamics" was held at the Research Institute for Mathematical Sciences of the Kyoto University from the 17th through the 19th of December, 1970, under the leadership of the research director, Prof. Minoru Urabe. At the seminar, I volunteered to make an oral report entitled "A steady solution of the nonlinear ordinary differential equations of the second order." The record of the following comments I made at the end of my research report still remains. "··· However, according to my observation of the phenomenon with the use of a computer, each of the minimal sets which make up the set of central points are all unstable, and the steady state seems to move randomly around the vicinity of the minimal sets, influenced by small fluctuations in the oscillatory system or external disturbances." [15] These minimal sets are of course the unstable periodic motions in the attractor; the above description led to my proposed name "randomly transitional phenomena." The reason I made the report in Prof. Urabe's seminar was because I hoped for the mathematicians to hear and possibly support my ultimate interpretation of the random oscillations. I was hoping especially because in those days Prof. Hayashi did not welcome my mention of "set of central points," or "minimal set," etc. during our seminars. Everytime my interpretation of the random oscillations was mentioned, he kept pressing me to examine the errors further. Despite my ardent hope, however, my gamble backfired that day. I cannot be sure of the date, since Prof. Urabe's seminar, although small in scale, was held every year. After my report, at any rate, Prof. Urabe admonished me personally. "What you saw was simply the essence of quasi-periodic oscillations." he said. "You are too young to make conceptual observations." The existence of random oscillations (chaos) was so obvious in my mind, that the negative comment did not crush me. Even so, I was deeply disappointed that no one understood it no matter how hard I tried to explain. From then on, I became even more careful in my research efforts.

I think it was 1971 when I sent off the second paper based on my Decem-

Strange Attractors and the Origin of Chaos

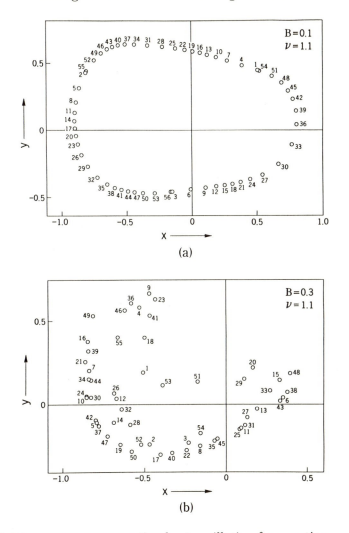

(a)

(b)

Fig. 15 Point sequences representing beat oscillation for equation

$$\dot{x} = y, \qquad \dot{y} = \mu(1 - \gamma y^2)y - x^3 + B\cos\nu t \qquad (8)$$

with $\mu = 0.2$, $\gamma = 4.0$, showing the difference between almost periodic oscillation and chaotic oscillation.

(a) $B = 0.1$, $\nu = 1.1$: invariant closed curve representing almost periodic oscillation.

(b) $B = 0.3$, $\nu = 1.1$: chaotic attractor.

(Reproduced with the courtesy of the Institute of Electronics and Communication Engineers of Japan [14])

207

ber 1970 oral presentation at the Research Institute for Mathematical Sciences mentioned above. Although I have very little record of it on hand, the Electronics and Communication Society must have some data. The paper was rejected. My only collaborator in random oscillation research, Akamatsu, had gone back to Tokushima, and I was all alone. I spent nearly a year rewriting it and sent it again on Sept. 7, 1972. After going through the review process, it was finally accepted [16]. The summary of the discrete dynamical system theory I included in the appendix was by far the most difficult work. I will never forget how nervous I was, wondering whether or not I had a full and accurate grasp of the concept. The mathematicians who valued rigorous proofs were, in a way, my bane. They can set up any unrealistic assumptions in their heads and live in their world of abstractions, but we are living in the real world. While I wanted them to hear me out a little more sympathetically, I also idolized mathematics. In the appendix, I summarized the essence of the papers by Birkhoff, Nemytskii and Stepanov, etc., but I simply could not understand the transfinite ordinal of the second ordinal class. As I reread this paper [16] now, I am a bit embarrassed by my poor writing. And yet this was where I advocated the existence of chaotically transitional oscillation. I feel that this paper holds true even today, except for a seemingly erroneous description of structural stability, and for the fact that the numerical examples in Figs. 8 and 9 obtained from the analog computer data, turned out to be non-chaotic, according to some later digital simulations which showed periodic oscillations instead. (These periodic oscillations are exceptional, and at typical nearby parameter values, both digital and analog simulations show chaotic oscillations.) Or rather, I should say my qualitative understanding of the steady chaotic phenomenon has not changed or advanced since then. Akamatsu came from Tokushima to write a clean copy of this paper for me. I had intentionally sent off the paper during Prof. Hayashi's absence, so that I had a good excuse for not having it reviewed by him, and it probably lacks certain fine editing because of it. But there was no way I could show the paper to Prof. Hayashi, since I knew he would make drastic changes and cut out what seemed essential to me. I could not compromise, however. It would have been more troublesome to show it to him and then clash with him than not show it at all. I was truly desperate. Even so, I had to consider the protocol of our research, as well as Akamatsu's position since he had not submitted his dissertation yet. Prof. Hayashi's name had to be included as a co-author. Ignoring such protocol may have been a lot simpler, but I could not do that. I have published several papers since then, and I always try to be very careful in selecting data. I would like to continue this practice in the future as well. I also try not to forget that there are junior colleagues who are plagued with insecurity but are clinging to the hope of a better future. It is regrettable that there are some people who are not ashamed to use them to their own advantage. If you have a chance to read Prof. Hayashi's paper on the Duffing equation [17] written around this time, the credibility of my account should be clear to you. In September of 1972, Prof. Hayashi was attending the Sixth International Conference on Nonlinear Oscillation held at Poznań. It sounds silly now, but I learned the hard way that changing the already established order in this world was truly a difficult task. Prof. Hayashi kindly disregarded this incident, but my paper [16] was not included in his **Selected Papers on Nonlinear Oscillations** published in 1975 [17]. The circumstance was such that I could not personally translate or proof-read the English version of the paper [16] published in **Scripta**.

7. THE END OF THE CHIHIRO HAYASHI LABORATORY

Prof. Hayashi retired in the spring of 1975. I was just an Associate Professor at that time and had no idea what had transpired. But the Hayashi Laboratory was dismantled several years later. I have never been clear on the reason for this. We were a group of lost souls without a leader, but I have fond memories of being able to do my work freely for the first time. In addition to the chaos research, I was studying nonlinear systems with time delay with Yoshiaki Inoue. I was also doing calculation of the power spectrum of chaos with the collaboration of the Institute of Plasma Physics, Nagoya University. After that I was invited into the laboratory of Prof. Chikasa Uenosono of Kyoto University where I have been allowed to stay, in the Engineering Dept., to this day. This period was also a chaotic time in my own life. In the spring of 1978, at the time of my move to the Uenosono Laboratory, my paper [18] was published in the **Journal of the Institute of Electrical Engineers**. This was the paper in which the strange attractor (Fig. 16) — the one for which Prof. David Ruelle coined the name "Japanese attractor" — was reported [19, 20]. What was new in the paper was the addition of the power spectrum of random oscillation and related properties to my earlier paper [16], but considering the five years between the two papers, I don't find much progress in it. As I was planning to present this paper to mark the end of my nonlinear oscillation research, I took special care in writing the whole paper. It was around the time that I first heard the term "chaos." But I had never imagined that the enormous amount of data on the Duffing equation system, etc. which I had accumulated, actually represented chaos itself, and that they would draw such wide attention later on.

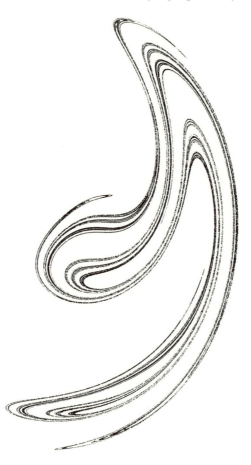

Fig. 16 Japanese attractor.

8. THE ORIGINAL DATA THAT WERE PRESERVED

My original data concerning the Japanese attractor, including the response curves of the shattered egg mentioned earlier, calculated by hand, and the output diagram of the analog computer, are preserved at present in the Brookhaven National Laboratory. In order to explain how they got there, I have to start with my meeting with Dr. Hugh Bruce Stewart.

It was June 1986 when I met Dr. Stewart for the first time. He had come all the way to Henniker, New Hampshire to see me, and attend SIAM's Conference on Qualitative Methods for the Analysis of Nonlinear Dynamics held under the leadership of F. M. A. Salam and M. L. Levi. For that entire week at Henniker, we had long talks. His interest was not limited to technical discussions. He was also interested in the origin of my chaos research. So I began with the shattered egg and reminisced about how it came about. On the last day of our meeting, he handed me a hand-written note which said, "I have gotten approval to acquire and preserve all of your chaos research material. Would you give it in its entirety to the Brookhaven National Laboratory? We will take the full responsibility for preserving it properly." He eventually typed up this letter and sent it to me in Japan. To me, the material had only some sentimental value, but it would have been worthless in Japan. Thinking that I would probably throw it away when leaving the University, I selected some relevant material and sent a boxful to Brookhaven. That gave me a good excuse to put in order my original data that had filled my office all these years. Except for a few things, the whole mess got cleared away, making much needed room in my office.

In April of 1988, Bruce came to Japan. While we were talking, he popped the question of when had I seen the Japanese attractor? I began telling him about the circumstances I had mentioned above, but he did not seem to be convinced. His usual warmth and friendliness gradually turned into a skeptical demeanor. I felt like I was being put on the spot. The day was April 11th, 1988. I fished out some files that I had not sent to BNL, and showed them to him. He studied the rumpled report papers thoroughly and checked my scribbled numbers with his calculator. Finally he seemed to be convinced, and returned to his usual self. I thought this was a good opportunity to clean up some more of the mess, and tossed the files into the wastebasket. But Mr. Bruce said, "Here, give them to me!" I wasn't particularly proud of my rumpled messy files, but agreed. He immediately attached a memo and sent them to Prof. Ralph Abraham at the University of California Santa Cruz. I have seen Bruce many times since then, but haven't seen that stern face ever again. He was probably thinking, "no matter what Yoshi says, I have to examine the actual evidence myself before deciding whether or not to believe it" — this must have been his credo, and I deeply respect that. Many of my results have been cited in his books and papers, but from what I heard, he decided to select them only after checking them carefully until he was absolutely convinced [21]. I believe that every scientist should practice such rigor.

9. EPILOGUE

What I have been telling you is nothing but my subjective account. Nevertheless, I intentionally mentioned real names, as I wanted this memoir to be accountable and traceable by the data behind it. Regrettably, Profs. Chihiro Hayashi and Minoru Urabe are no longer with us and won't be able to dispute my account. In order to keep the context intact, however, I could not avoid telling this obviously one-sided story.

As for Prof. Chihiro Hayashi, for sixteen long years, from my graduate student days to the day of his retirement, I received his guidance. He was the last of the true breed of Meiji Era imperial university professors in the Electrical Engineering Dept. of the School of Engineering, Kyoto University. I tried to tone down the description of his personality, but as you can see in this account, he had a very strong personality. He was the emperor of his laboratory, and yet outwardly he was a mild-mannered gentleman. I believed at that time that his was the most feudalistic of any laboratory in the world, and the wall of his authority was impenetrable during his reign, and I still believe it. But because of that, we were not swept away by worldly concerns and could concentrate on our research, being faithful to our own ideas. For this I am truly thankful. It is obvious that my research was not possible without Prof. Hayashi's presence. The rigor, will, and courage needed for one's own research — I trust that these were Prof. Hayashi's legacy. However, I was hounded by his obsession for cleanliness in diagrams.

Later I received guidance from Prof. Chikasa Uenosono. Under his guidance, I learned the rigor and intensity of the world of technology. At the same time I had an opportunity to reaffirm my belief in the importance of experiments as well as accurate perception of phenomena. Prof. Uenosono also taught me the rules and order of the human world and how to handle them. These things have greatly enriched my life. Furthermore, I would not have been sitting in my present position, had it not been for Prof. Uenosono. Needless to say, I am deeply indebted to these two professors. However, the greatest influence came from Prof. Michiyoshi Kuwahara, to whom I am most grateful .

As I have focused this account on my chaos research alone, you may have had the impression that I reached the idea of randomly transitional phenomena directly without going astray. In reality, however, it was a long, meandering and groping process, as you can see if you draw the time line of my career.

What I have been working on during the period I described, can be called research in nonlinear oscillation, or from a larger point of view, in basic electrical engineering or applied mathematics. I have a feeling that the people who are involved in these fields are more or less criticized continually by both mathematicians and engineers just the way I have been. The criticisms can be summarized as follows: "Has it been proven?" "Is it useful in some way?" These are reasonable questions indeed, and I have always been at a loss for an answer.

I would like to take this opportunity to bring forth a counter-argument. To mathematicians, I would like to ask "How sound is your proposition?" and to engineers I would like to recommend a reading of Bob Johnstone's article entitled "Research and Innovation: No chaos in the classroom" in the **Far Eastern Economic Review** [22]. In the article he quotes Tien-Yien Li's observation (Li of the Li-Yorke's theorem), "If the applicability of chaos becomes apparent, the Japanese will show a fierce interest in it."

Whatever the case may be, I believe it is most important, especially for Japanese researchers, to make an unbiased evaluation of one another's positions based on accurate communication. In this sense, the role of this symposium is truly momentous.

REFERENCES

1. C. Hayashi, H. Shibayama and Y. Ueda, Quasi-periodic oscillations in self-oscillatory systems with external force (in Japanese), *IECE Technical Report*, Nonlinear Theory (Dec. 16, 1961).

2. C. Hayashi, *Nonlinear Oscillations in Physical Systems*. McGraw-Hill, New York (1964): Reissue, Princeton Univ. Press (1984).

3. C. Hayashi and Y. Nishikawa, Initial conditions leading to different types of periodic solutions for Duffing's equation, *Proc. Int. Symp. Nonlinear Oscillations, Kiev*, Vol. 2, pp. 377-393 (1963).

4. C. Hayashi, *Forced Oscillations in Nonlinear Systems*. Nippon Printing and Publishing Co., Osaka, Japan (1953).

5. C. Hayashi, Y. Nishikawa and Y. Ueda, Higher-harmonic oscillations in nonlinear circuits (in Japanese), *IECE Technical Report*, Nonlinear Theory (Nov. 29, 1963).

6. C. Hayashi, Higher harmonic oscillations in nonlinear forced systems, *Colloq. Int. CNRS, Marseilles*, No. 148, pp. 267-285 (1965).

7. K. W. Blair and W. S. Loud, Periodic solutions of $x'' + cx' + g(x) = Ef(t)$ under variation of certain parameters, *J. Soc. Ind. Appl. Math.* **8**, 74-101 (1960).

8. N. Levinson, Transformation theory of non-linear differential equations of the second order, *Ann. Math.* **45**, 723-737 (1944).

9. C. Hayashi, Y. Ueda and H. Kawakami, Transformation theory as applied to the solutions of non-linear differential equations of the second order, *Int. J. Non-Linear Mech.* **4**, 235-255 (1969).

10. C. Hayashi, Y. Ueda, H. Kawakami and S. Shirai, Analysis of Duffing's equation by using mapping procedure (2) (in Japanese), *IECE Technical Report*, NLP67-13 (Dec. 8, 1967).

11. T. Endo and T. Saito, Chaos in electrical and electronic circuits and systems, *Trans. IEICE* **73-E**, 763-771 (1990).

12. C. Hayashi, Y. Ueda and H. Kawakami, Solution of Duffing's equation using mapping concepts, *Proc. Fourth Int. Conf. on Nonlinear Oscillations, Prague*, pp. 25-40 (1968).

13. C. Hayashi and Y. Ueda, Behavior of solutions for certain types of nonlinear differential equations of the second order, *Proc. Sixth Int. Conf. on Nonlinear Oscillations, Poznań*, Vol. 14, pp. 341-351 (1973).

14. C. Hayashi, Y. Ueda, N. Akamatsu and H. Itakura, On the behavior of self-oscillatory systems with external force (in Japanese), *Trans. IECE Japan* **53-A**, 150-158 (1970).

15. C. Hayashi, Y. Ueda and N. Akamatsu, On Steady-state solutions of a nonlinear differential equation of the second order (in Japanese), *Research Report RIMS, Kyoto University*, No. 113, pp. 1-27 (1971).

16. Y. Ueda, N. Akamatsu and C. Hayashi, Computer simulation of nonlinear differential equations and non-periodic oscillations (in Japanese), *Trans. IECE Japan* **56-A**, 218-225 (1973).

17. C. Hayashi, *Selected Papers on Nonlinear Oscillations*. Nippon Printing and Publishing Co., Osaka, Japan (1975).

18. Y. Ueda, Random phenomena resulting from nonlinearity – in the system described by Duffing's equation (in Japanese), *Trans. IEE Japan* **98-A**, 167-173 (1978).

19. D. Ruelle, Les attracteurs étranges, *La Recherche* **11**, 132-144 (1980).

20. D. Ruelle, Strange Attractors, *The Mathematical Intelligencer* **2**, 126-137 (1980).

21. J. M. T. Thompson and H. B. Stewart, *Nonlinear Dynamics and Chaos*. John Wiley and Sons, Chichester (1986).

22. B. Johnstone, No chaos in the classroom, *Far Eastern Economic Review*, p. 55, (22 June 1989).

23. Y. Ueda, Steady motions exhibited by Duffing's equation: a picture book of regular and chaotic motions, *New Approaches to Nonlinear Problems in Dynamics*, edited by P. J. Holmes, pp. 311-322, SIAM, Philadelphia (1980).

24. J. Gleick, *CHAOS: Making a New Science*. Viking Penguin Inc. (1987).

APPENDIX: Circumstances of the Publication of This Article in "Nonlinear Science Today" Vol. 2, No. 2 (1992)

I have known Philip Holmes, the Managing Editor of **Nonlinear Science Today**, since the summer of 1978, when my friend, Kazutaka Makino, sent Prof. Holmes reprints of some of my papers through Prof. Jenkins, a friend of Dr. Makino's. Our first meetings took place in December of 1979, when I was invited to a conference entitled "New Approaches to Nonlinear Problems in Dynamics" organized by Prof. Holmes and held at the Asilomar Conference Grounds, in Monterey Peninsula, California. This was the first time my work on chaos received international recognition, which was a truly memorable occasion for me [23]. This led to several more meetings between the two of us. While taking a leisurely stroll in Nara during his visit to Japan in the summer of 1990, Prof. Holmes asked me whether or not I was interested in writing a memoir of my work in chaos for an upcoming issue of **Nonlinear Science Today**. "I have been toying with the idea of writing such a memoir," I told him, "but right now I have no time. If you can wait for a few years, I would be most happy to try." Then came the invitation from the UN University to speak on that very subject. I regretted submitting my memoir to the UN University's symposium before I could do so for Phil Holmes, but I knew that the opportunity might not arise again. After the symposium the copy of the speech was sent to Phil with my apology. Phil's response was one of delight that the memoir was written at all, whatever the circumstances. I, therefore wish to note here the circumstances under which this memoir is finally reaching the readers of this magazine, and to express my deep gratitude to Professor Philip John Holmes for his generosity and patience.

ACKNOWLEDGMENTS

This article originally written in Japanese was translated by Mrs. Masako Ohnuki. The sequence of the events is as follows. The book **"CHAOS: Making a New Science"** by James Gleick was published in 1987 by **Viking Penguin Inc.** [24]. This book was very popular and stayed on the New York Times best-seller list for more than half a year, and was widely admired for its skillful explanations. The present author received a letter from Mr. James Gleick when this book was to be translated into Japanese by his nominated translator, Mrs. Masako Ohnuki. He requested me to supervise her translation. When the author read her translation, he was completely struck with admiration. The translation is accurate and there is no sense of incompatibility peculiar to a translation.

The author lacks the skill to write on delicate matters in English, and he asked her to translate the article, and she kindly agreed. He also requested

Dr. Hugh Bruce Stewart to review her translation and to check the contents of this article for historical accuracy. The author would like to express his sincere thanks to Mrs. Masako Ohnuki and Dr. Hugh Bruce Stewart. He also expresses his thanks to the following colleagues:

Chihiro Hayashi	deceased
Minoru Urabe	deceased
Chikasa Uenosono	Professor Emeritus, Kyoto University
Hiroshi Shibayama	Professor Emeritus, Osaka Institute of Technology
Michiyoshi Kuwahara	Professor Emeritus, Kyoto University
Yoshikazu Nishikawa	Kyoto University
Minoru Abe	Kyoto University
Nobuo Sannomiya	Kyoto Institute of Technology
Masami Kuramitsu	Kyoto University
Hiroshi Kawakami	Tokushima University
Norio Akamatsu	Tokushima University

LIST OF PUBLICATIONS IN ENGLISH
ON NONLINEAR DYNAMICS

1. C. Hayashi, H. Shibayama and Y. Ueda, Quasi-periodic oscillations in a self-oscillatory system with external force, *Proc. Symp. Nonlinear Oscillations (Intern. Union Theoret. Appl. Mech. Kiev)*, Vol. 1, pp. 495-509 (1963).

2. C. Hayashi and Y. Ueda, Forced negative resistance oscillator, *Proc. Int. Conf. Microwaves, Circuit Theory and Information Theory, Tokyo*, Part 2, pp. 107-108 (1964).

3. Y. Ueda, *Some Problems in the Theory of Nonlinear Oscillations*. Doctoral Dissertation, submitted to the Faculty of Engineering, Kyoto University, February 1965.

4. C. Hayashi, Y. Ueda and H. Kawakami, Solution of Duffing's equation using mapping concept, *Proc. Fourth Conf. Nonlinear Oscillations, Prague*, pp. 25-40 (1968).

5. Y. Ueda, *Some Problems in the Theory of Nonlinear Oscillations*. Nippon Printing and Publishing Co., Osaka, Japan (1968).

6. C. Hayashi, Y. Ueda and H. Kawakami, Transformation theory as applied to the solutions of non-linear differential equations of the second order, *Int. J. Non-Linear Mechanics* **4**, 235-255 (1969).

7. C. Hayashi, Y. Ueda, N. Akamatsu and H. Itakura, On the behavior of self-oscillatory systems with external force, *Trans. IECE Japan* **53-A**, 150-158 (1970): English translation, *Electronics and Communications in Japan*, pp. 31-39, Scripta Publ. Co., Silver Spring, MD.

8. C. Hayashi, Y. Ueda and H. Kawakami, Periodic solutions of Duffing's equation with reference to doubly asymptotic solutions, *Proc. Fifth Int. Conf. Nonlinear Oscillations, Kiev*, Vol. 2, pp. 507-521 (1970).

9. C. Hayashi and Y. Ueda, Behavior of solutions for certain types of nonlinear differential equations of the second order, *Nonlinear Vibration Problems (Proc. Sixth Int. Conf. Nonlinear Oscillations, Poznań)*, Vol. 14, pp. 341-351 (1973).

10. Y. Ueda, N. Akamatsu and C. Hayashi, Computer simulation of nonlinear ordinary differential equations and non-periodic oscillations, *Trans. IECE Japan* **56-A**, 218-225 (1973): English translation, *Electronics and Communications in Japan*, pp. 27-34, Scripta Publ. Co., Silver Spring, MD.

11. Y. Ueda and Y. Inoue, Forced oscillations in a nonlinear system with time delay, *Trans. IEE Japan* **95-A**, 239-246 (1975): English translation, *Electrical Engineering in Japan*, pp. 34-42, Scripta Publ. Co., Silver Spring, MD.

12. Y. Inoue and Y. Ueda, Two-frequency almost periodic oscillations in a nonlinear forced system with time delay, *Trans. IEE Japan* **96-A**, 441-448 (1976): English translation, *Electrical Engineering in Japan*, pp. 9-16, Scripta Publ. Co., Silver Spring, MD.

13. C. Hayashi and Y. Ueda, The method of mapping as applied to the solution of nonlinear differential equations: with reference to doubly asymptotic solutions, *Proc. Seventh Int. Conf. Nonlinear Oscillations, Berlin* (1975).

14. Y. Ueda, Random phenomena resulting from nonlinearity: in the system described by Duffing's equation, *Int. J. Non-Linear Mechanics* **20**, 481-491 (1985): Translated from *Trans. IEE Japan* **98-A**, 167-173 (1978).

15. Y. Ueda, Randomly transitional phenomena in the system governed by Duffing's equation, *J. Statistical Physics* **20**, 181-196 (1979).

16. Y. Ueda, Steady motions exhibited by Duffing's equation: a picture book of regular and chaotic motions, *New Approaches to Nonlinear Problems in Dynamics*, edited by P. J. Holmes, pp. 311-322, SIAM (1980).

17. Y. Ueda, Explosion of strange attractors exhibited by Duffing's equation, *Nonlinear Dynamics* (*Annals of the New York Academy of Sciences* **357**), pp. 422-434 (1980).

18. Y. Ueda and N. Akamatsu, Chaotically transitional phenomena in the forced negative-resistance oscillator, *Trans. IEEE* **CAS-28**, 217-224 (1981).

19. Y. Ueda, Self-excited oscillations and their bifurcations in systems described by nonlinear differential-difference equations, *Proc. 24th Midwest Symposium on Circuits and Systems*, edited by S. Karni, pp. 549-553 (1981).

20. H. Ogura, Y. Ueda and Y. Yoshida, Periodic stationality of a chaotic motion in the system governed by Duffing's equation, *Prog. Theor. Phys.* **66**, 2280-2283 (1981).

21. Y. Ueda and H. Ohta, Strange attractors in a system described by nonlinear differential-difference equation, *Chaos and Statistical Methods*, edited by Y. Kuramoto, pp. 161-166, Springer-Verlag (1984).

22. Y. Ueda and H. Ohta, Average power spectra of chaotic motions in a system described by nonlinear differential-difference equation, *Proc. ISCAS 85*, pp. 179-182 (1985).

23. Y. Ueda, Survey of strange attractors and chaotically transitional phenomena in the system governed by Duffing's equation, *Complex and Distributed Systems: Analysis, Simulation and Control*, edited by S. G. Tzafestas and P. Borne, pp. 173-180, Elsevier Science Publishers B. V. (1986).

24. Y. Ueda, H. Nakajima, T. Hikihara and H. B. Stewart, Forced two-well potential Duffing's oscillator, *Dynamical Systems Approaches to Nonlinear Problems in Systems and Circuits*, edited by F. M. A. Salam and M. L. Levi, pp. 128-137, SIAM (1988).

25. Y. Ueda and S. Yoshida, Attractor-basin phase portraits of the forced Duffing's oscillator, *Proc. Euro. Conf. Circuit Theory and Design, Paris*, Vol. 1, pp. 281-286 (1987).

26. J. M. T. Thompson and Y. Ueda, Basin boundary metamorphoses in the canonical escape equation, *Dynamics and Stability of Systems* **4**, 285-294 (1989).

27. Y. Ueda, S. Yoshida, H. B. Stewart and J. M. T. Thompson, Basin explosions and escape phenomena in the twin-well Duffing oscillator: compound global bifurcations organizing behaviour, *Phil. Trans. R. Soc. Lond.* **332-A**, 169-186 (1990).

28. Y. Ueda, Survey of regular and chaotic phenomena in the forced Duffing oscillator, *Int. J. Chaos, Solitons and Fractals* **1**, 199-231 (1991).

29. H. B. Stewart and Y. Ueda, Catastrophes with indeterminate outcome, *Proc. R. Soc. Lond.* **432-A**, 113-123 (1991).

30. H. B. Stewart, J. M. T. Thompson, A. N. Lansbury and Y. Ueda, Generic patterns of bifurcation governing escape from potential wells, *Int. J. Bifurcation and Chaos* **1**, 265-267 (1991).

31. Y. Ueda, T. Enomoto and H. B. Stewart, Chaotic transients and fractal structures governing coupled swing dynamics, *Applied Chaos*, edited by Jong Hyun Kim and John Stringer, pp. 207-218, John Wiley and Sons, Chichester (1992).

32. Y. Ueda, Strange attractors and the origin of chaos, *Int. J. Nonlinear Science Today*, Vol. 2, No. 2, pp. 1-16 (1992); will also appear in the *Proc. Int. Symp. "The Impact of Chaos on Science and Society,"* April 15-17, 1991 organized by the United Nations University and the University of Tokyo.

POSTSCRIPT

I realize that leaving one's own writing behind may be like leaving one's inade-
quacies behind. Yet it would still be nice to be able to leave some mark of my
existence in this world, or rather, to be rid of my inferiority complex for not
penning one single book after being a university educator and researcher for so
long. It is true, I have written numerous research papers, but at this time of
my life, I never even dreamed of publishing a collection of my papers.

The publication of this book was made possible by the suggestion and planning of Professor Ralph Abraham of the University of California, Santa Cruz,
and by the generous assistance of Dr. Bruce Stewart of the Brookhaven National Laboratory. It was November of 1991 when Prof. Abraham informed
me of his desire to publish a collection of the reprints of my papers written in
English. The first thing that came to my mind at that time was my limited
English. Whenever I needed to write papers in English, I first made a pile in
front of me of English textbooks and papers on the pertinent subject. Then I
fished around for the sentences that seemed to say what I wanted to say (this
is true to this day). It took a great amount of time. Though frustrated when
the borrowed sentences were not quite right, I often had to make compromises.
Now that they are going to be published, it seemed to me to be an opportune
time to reexamine my English writings, at least those with no formal reprints
(Selections 1 through 3). When asked to take on this task, Dr. Bruce Stewart,
with whom I am presently conducting collaborative research, graciously agreed,
and has gone through them not only once but twice. I requested that the reex-
amination be limited only to my English expressions, and that the contents of
the papers be left at the level of my understanding of the subject at the time
they were written. He followed this request to the letter, but somehow came
up with a page of additions. They were truly reasonable additions, but my
stubborn principles won, and I had to remove some parts of them without his
knowledge.

I have appended a list of my achievements at the end of this book; six
of the papers (Selections 1 through 6) in the list, and a memoir (Selection 7)
reminiscing on those days when I was writing the papers are included in the
book. As the readers can see from Selection 7, I was quite defiant at that
time, believing in my honest observations with little regard to other people's
opinions. I firmly believed that what I saw was not quasi-periodicity at all,
and that my papers should be nothing other than the product of my belief in

the unadulterated observations I made. I felt I could not help it if they seemed erroneous to others, especially to those who had never actually done the work themselves. My credo was that the true worth of research is not decided by majority rule. It was truly a stroke of luck that the results of my research finally met acceptance when chaos garnered the center stage in the late 1970s. Since 1980, however, my work has been mostly supplementary clarification of my original observations. Thanks to Drs. Abraham and Stewart, the publication of my works will bring to a conclusion my past 55 years, and will give me an opportunity to start afresh.

I would like to express my deepest gratitude to Professor Abraham and Dr. Stewart who gave this book a wonderful title, led me through this project with great patience, and made it all come true. I would also like to thank my wife, Miyoko, for her patience and understanding through these years.

Yoshisuke Ueda
Kyoto, Japan

CURRICULUM VITAE OF YOSHISUKE UEDA

December 23, 1936	Born in Kobe City
1959	Graduated Kyoto University, B.E. degree in Electrical Engineering
1964	Finished Graduate School of Engineering, Kyoto University
1965	D.E. degree from the Faculty of Engineering, Kyoto University
1964-1967	Instructor of Electrical Engineering, Kyoto University
1967-1971	Lecturer of Electrical Engineering, Kyoto University
1971-1985	Associate Professor of Electrical Engineering, Kyoto University
1982-1984	Chairman of the Professional Group on Nonlinear Problems, Institute of Electronics, Information and Communication Engineers of Japan
1985-	Professor of Electrical Engineering, Kyoto University
1989-1992	Chairman of the Professional Group on Rotating Machines, Institute of Electrical Engineers of Japan
1991-	Member of the Advisory Board of an Interdisciplinary Journal of Nonlinear Science: CHAOS (American Institute of Physics)
1991-	Member of the Editorial Board of the International Journal of Bifurcation and Chaos
1991-	Member of the Honorary Editors of the International Journal of Chaos, Solitons and Fractals
1992-	Member of the board of directors, Institute of Electrical Engineers of Japan